BIOINFORMATICS

BIOINFORMATICS

By

ARCHANA VERMA
MIS (Australia), M.C.A., M.Tech. (CSE)
Associate Professor,
Department of Information Technology
IEC College of Engineering and Technology
Greater Noida (U.P.)

UNIVERSITY SCIENCE PRESS
(An Imprint of Laxmi Publications Pvt. Ltd.)
An ISO 9001:2008 Company
BENGALURU • CHENNAI • COCHIN • GUWAHATI • HYDERABAD
JALANDHAR • KOLKATA • LUCKNOW • MUMBAI • RANCHI • NEW DELHI
BOSTON (USA) • ACCRA (GHANA) • NAIROBI (KENYA)

BIOINFORMATICS

Printed and bound in India
Typeset at Kalyani Computer Services, Delhi
First Edition : 2016
ISBN 978-93-85750-79-3

Branches

✆	Bengaluru	080-26 75 69 30	
✆	Chennai	044-24 34 47 26,	24 35 95 07
✆	Cochin	0484-237 70 04,	405 13 03
✆	Guwahati	0361-254 36 69,	251 38 81
✆	Hyderabad	040-27 55 53 83,	27 55 53 93
✆	Jalandhar	0181-222 12 72	
✆	Kolkata	033-22 27 43 84	
✆	Lucknow	0522-220 99 16	
✆	Mumbai	022-24 91 54 15,	24 92 78 69
✆	Ranchi	0651-220 44 64	

PUBLISHED IN INDIA BY

UNIVERSITY SCIENCE PRESS
(An Imprint of Laxmi Publications Pvt. Ltd.)

An ISO 9001:2008 Company
113, GOLDEN HOUSE, DARYAGANJ,
NEW DELHI - 110002, INDIA
Telephone : 91-11-4353 2500, 4353 2501
Fax : 91-11-2325 2572, 4353 2528
www.laxmipublications.com info@laxmipublications.com

C—11200/015/12
Printed at : Repro Knowledgecast Limited, Thane

Dedicated To

My Inspiring Father
Sh. Vidya Sagar Verma

My Supportive Mother
Smt. Sudha Verma

My Loving Sister
Dr. Vandana Verma Marwah

My Encouraging Brother
Mr. Shashank Verma

My Sincere Brother-in-Law
Dr. Ashutosh Marwah

My Sweet Sister-in-Law
Mrs. Archana Devar Verma

My Beloved Daughter
Ritambhara

My Beloved Niece
Aparajita

My Beloved Nephew
Shreyas

CONTENTS

PREFACE

I was inspired to write this book when the students appreciated my notes and found them useful for preparing for their exams. I realized that various topics were explained properly in different books and the students had difficulty in purchasing so many different books. So I thought of putting the best explanations together for their convenience and easy guidance.

The book is primarily meant for B.Tech Final Year students of U.P. Technical University, Lucknow, but I believe it will benefit students of other universities also, studying the same course.

The topics have been explained with suitable examples and diagrams wherever necessary.

The book contains an Introductory chapter on Biology for those who have not studied Biology earlier, supplemented with Introduction to Bioinformatics, Molecular Biology Data, Nucleic Acids, Protein—Structure, Folding and Function, Nucleotide Sequence Data, Biological Data Types and the Role of IT in Bioinformatics.

I do hope that this book will be useful for the students. I would feel highly rewarded if the students are benefitted by this book.

—Author

Acknowledgement

I deeply thank the Chairman, IEC Group of Institutions, Sh. R.L. Gupta for his blessings and encouragement for writing this book. I also take this opportunity to thank the CEO of IEC Group of Institutions, Sh. Naveen Gupta for his encouragement and support.

I sincerely thank the Director, IEC Group of Institutions, Prof (Dr.) D.B. Singh for his guidance and good wishes extended towards me.

I would like to thank the Chairman of Laxmi Publications Pvt. Ltd. Sh. Rajendra Kr. Gupta and Managing Director Sh. Saurabh Gupta and the staff members for publishing the book.

Lastly, I would like to thank my students for their appreciation of my notes which encouraged me to write this book.

—Author

Unit 1

INTRODUCTION TO BIOLOGY

1.1 INTRODUCTION TO THE CELL

It is the cell that makes an organism living in all living organisms. Each and every organism is made up of cells, either single celled or multiple celled. The organisms which have a single cell in them are called *unicellular organisms* and those organisms which are made up of many cells are called *multi-cellular organisms.*

Unicellular organisms are capable of two things, firstly of independent existence and secondly of performing the essential functions of life. Therefore, the cell is the one unit which makes up the structure of the living organism and does all the functioning of the living organism as well.

Scientist had observed many animal cells and found out that a thin outer layer was present as the cover of the cells and called it the *plasma membrane. Schwann* had said that cells were the basic units which made the animal and plant bodies and some cell products also helped in making those bodies.

Later on, it was known as the *cell theory* which is stated as under:

(1) "All living organisms are composed of cells and cell products."

(2) "All cells arise from pre-existing cells."

The cell has an outer membrane which is its delimiting structure. Inside each cell, lies a structure which has a membrane and is dense in nature and it is known as the *nucleus.* The nucleus has chromosomes inside it and the chromosomes then contain the genetic material more commonly referred to as the DNA.

The prokaryotic cells have a nucleus which is not bound by any membrane whereas the eukaryotic cells have a nucleus which is bound by a membrane. Inside the cell beside other organelles most of the volume is occupied by a half-fluid complex known as the *cytoplasm.* The cytoplasm is the area where all the cellular activities take place in both plant and animal cells. By cellular activity we mean the chemical reactions taking place which make the cell's state living. Without the reactions the cell would end up in a dead state.

Inside the cell, apart from the cytoplasm and the nucleus, many other structures which are also known as the organelles are found. They are:

1. The Endoplasmic Reticulum
2. The Golgi Apparatus
3. Lysosomes
4. Mitochondria
5. Microbodies
6. Vacuoles
7. Ribosomes

The ribosomes are another type of organelles which also do not have a membrane and are present in prokaryotic cells and eukaryotic cells. In prokaryotic cells, they are found inside the cytoplasm and in eukaryotic cells they are found not only inside the cytoplasm but attached to other organelles such as mitochondria and rough endoplasmic reticulum. Centriole is one of the other organelles which does not have a membrane and their purpose is to help in cell division.

1.2 PROKARYOTIC CELL

The prokaryotic cells are of the organisms such as bacteria, blue green algae, mycoplasma and PPLO (pleuro pneumonia like organisms).

The prokaryotes have two walls delimiting the cell known as the cell wall which surrounds the cell membrane.

The nucleus is not defined properly and most of the cell volume is filled with the cytoplasm. The genetic material is not surrounded by the nuclear membrane. The genomic DNA is in the form of a circular DNA or single chromosome. Apart from this, some small circular DNAs are found in addition to the genomic DNA.

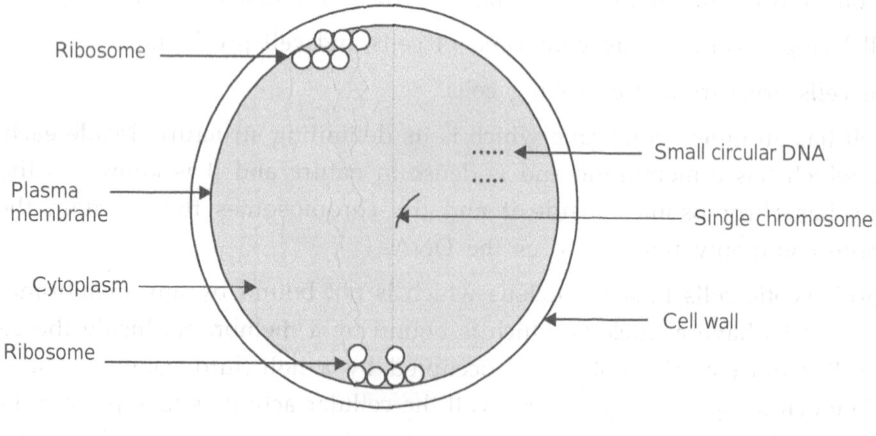

Fig. 1.1: The Cell

Nuclear membrane is found only in eukaryotes and ribosomes are the only organelles which are found in eukaryotes and prokaryotes.

In prokaryotes as well as eukaryotes, the cell is protected by a three layer unit which consists of glycocalyx, the cell wall and the plasma membrane. The glycocalyx is the outermost layer, then comes the cell wall and the inner most layer is the plasma membrane. The function of the cell wall is to provide shape and structural support to the cell so that the cell does not burst or collapse. The role of the plasma membrane is to contact the outside environment. It is also semipermeable so that material can flow from inside to outside and outside to inside.

The ribosomes in prokaryotes are mostly attached to the plasma membrane of the cell. The ribosomes are around 15nm to 20nm long and are formed of two subunits 40S and 30S which form the 70S prokaryotic ribosomes. The 40S is the large ribosomal unit and the 30S is the small ribosomal unit. Ribosomes are the site of protein synthesis (that is where the proteins are made). Many ribosomes are connected to one mRNA (messenger RNA) and make up a sequence which is known as polyribosomes or polysome.

1.3 EUKARYOTIC CELL

The eukaryotes are of the type of plants, animals and fungi. The most distinguishing feature of the eukaryotes is that the cytoplasm is separated from the other organelles by the presence of membranes bound to those organelles. Also the nucleus is membrane bound by a nuclear membrane. The genetic material of the eukaryotes is organized into chromosomes.

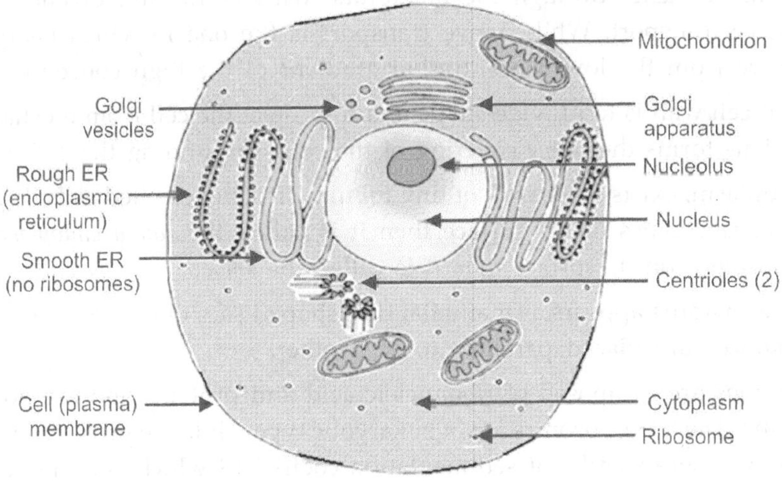

Fig. 1.2: The Eukaryotic Cell

The cell membrane is made up of lipids that are placed in a bi-layer fashion, also it contains proteins and carbohydrates. In human beings, the membrane has 52% proteins and 40% lipids. The proteins of the membrane are either known as integral or peripheral.

Proteins that are on the periphery or the upper part are present on the outside of the membrane and are known as the peripheral proteins, and the proteins which are internal are present in the inner core of the membrane and they are called the integral proteins.

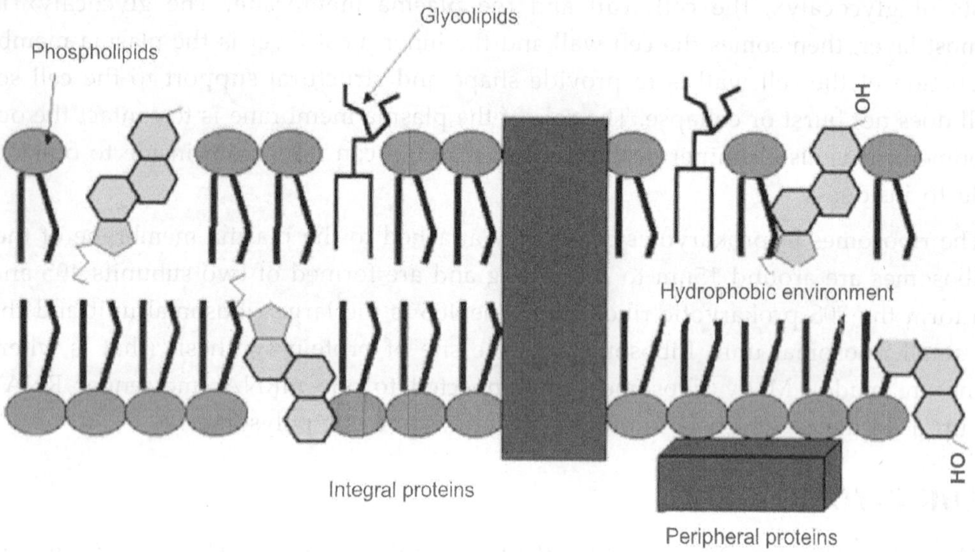

Fig. 1.3: The Plasma Membrane

The contribution of the plasma membrane is to transport the molecules across it. The membrane is able to leak a number of molecules which are present on any side of it. Many molecules can move easily through the membrane without the use of energy and this is called the passive transport. While active transport is the one in which energy is utilized that is ions move from the low concentration gradient to the high concentration gradient.

The role of cell wall is to provide shape and to protect the cell from mechanical damage and infection. This forms the outer covering of the cell surrounding the plasma membrane.

In the cytoplasm, exists a network of tiny tubular structures called *endoplasmic reticulum*. If the ER shows ribosomes on its surface then it is called the *rough endoplasmic reticulum*, if it has no ribosomes on its surface then it is called the *smooth endoplasmic reticulum.*

The golgi apparatus appears as many flat disc shaped sacs of 0.5 μm to 1.0 μm diameter. The golgi apparatus are placed parallely to each other.

The ribosomes are composed of ribonucleic acid and proteins and are not surrounded by any membrane. The 80S ribosomes are of eukaryotic type while the prokaryotic counterparts are 70S. The S over here stands for sedimentation coefficient which is a measure of density and size for the ribosomes. The 70S and 80S ribosomes are made up of two subunits. In eukaryotic ribosomes, the large ribosomal unit is of 50S and the small ribosomal unit is of 30S.

The centrosome is an organelle containing two cylindrical structures called *centrioles*.

The 2 centrioles are placed perpendicularly to each other in the centrosome and each of the centrioles has a cartwheel type of organization. The centriole's central part is also made up of proteins and is known as the hub. It is connected with three small peripheral tubes in radial straight lines made of proteins.

When the nucleus is dividing, it is known as the metaphase nucleus and when the nucleus is not dividing it is known as the interphase nucleus.

In the interphase nucleus, the nucleus is arranged into very comprehensive and detailed fibres made up of nucleoprotein which are known as chromatin and the nuclear matrix which is commonly known as the *nucleoplasm* is set inside it.

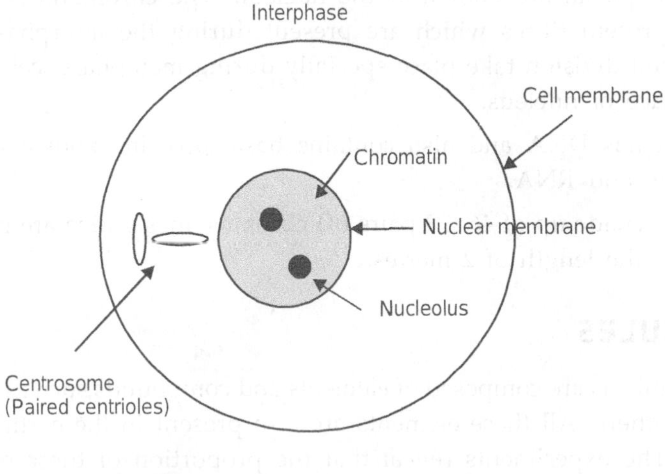

Fig. 1.4: The Interface Nucleus

During metaphase the nucleus has well defined chromosomes which contain the DNA, the nucleolus and the nucleoplasm.

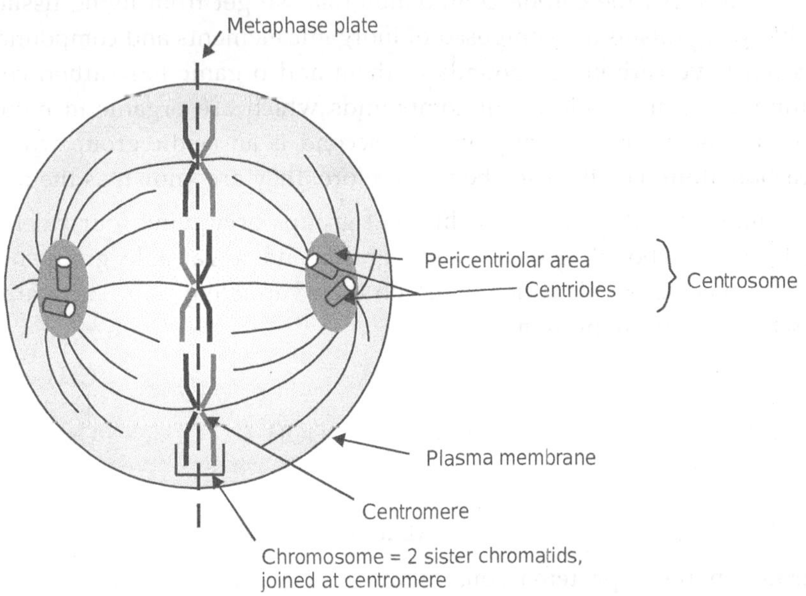

Fig. 1.5: The Nucleus During Metaphase

There are 2 parallel membranes covering the nuclear envelope to prevent materials to escape into the cytoplasm. The space between the 2 parallel membranes is known as the perinuclear space and is around 10 to 50 nm wide. It lets the materials stay inside the nucleus. At distant intervals the nuclear envelope bears pores which are formed by fusion of its two membranes. The nuclear pores allow the passage of RNA and protein molecules outside to the cytoplasm and also in the opposite direction, i.e., from the cytoplasm to the nucleoplasm.

The nucleoplasm is made up of chromatin fibres and the nucleolus. Spherical structures present in the nucleoplasm are known as the nucleoli. The chromatin is a loose indistinct network of nucleoprotein fibres which are present during the interphase nucleus. When different stages of cell division take place specially during metaphase, cells show structured chromosomes in place of nucleus.

The chromatin has DNA and also contains basic proteins known as histones, some non-histone proteins and RNA.

A human cell is made up of 46 (23 pairs of) chromosomes which are present in a strand of DNA which is of the length of 2 metres.

1.4 BIOMOLECULES

The living organisms are composed of elements and compounds such as carbon, hydrogen, oxygen and many others. All these elements are also present in the earth's crust as well as in a living tissue. The experiments reveal that the proportion of these elements' presence is more in the living organisms than in the earth's crust.

If we take a living tissue and grind it in trichloroacetic acid, we obtain the acid soluble portion and the acid insoluble portion. In this acid—soluble pool, thousands of organic compounds are found. All the carbon compounds that we get from living tissues are called *biomolecules.* Living organisms are composed of inorganic elements and compounds, inorganic elements does not have carbon compounds in them and organic has carbon compounds in them. The amino acids are made up of compounds which are organic in nature and they have 2 groups, first is an amino group and the second is an acidic group which take place on the same carbon atom, i.e., the α-carbon. Therefore, they are known as the α-amino acids.

In the α-amino acids there are four different groups occupying four valency positions. These are hydrogen, carboxyl, group, amino group and a variable group designated as R-group. The amino acids are identified based on their variable group. There are 20 different amino acids which occur in proteins.

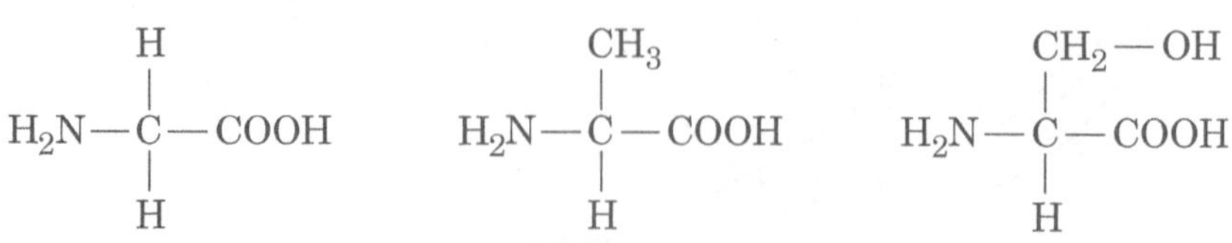

Glycine Alanine Serine

The R-group in these proteinaceous amino acids is a hydrogen (glycine), a methyl group (alanine), hydroxyl methyl (serine) etc.

The amino and the carboxyl groups determine whether the amino acids are acidic (glutamic acid), basic (lysine) or neutral (valine). There are some aromatic amino acids such as tyrosine, phenylaline and tryptophan.

The amino acids are able to form ions and interact with other groups so their structure changes. This is the property of NH_2 and COOH groups.

Lipids are generally water insoluble, normally a fatty acid. A fatty acid has a carboxyl group attached to an R-group. The R-group is either a methyl (CH_3) or ethyl (C_2H_5) or any higher number of CH_2.

There are a number of carbon compounds which contain heterocyclic rings in living organisms. Some of them contain nitrogen bases and are known as adenine, guanine, cytosine, uracil and thymine. The nitrogen bases attached to a sugar are called nucleosides and if a phosphate group is attached to the nucleosides it is called a *nucleotide.*

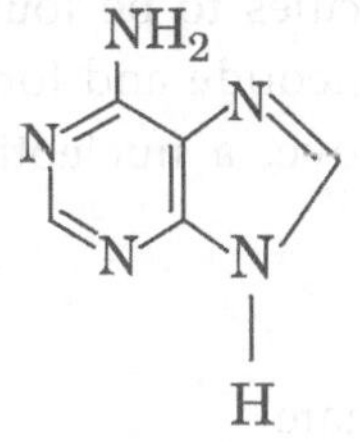

Adenine

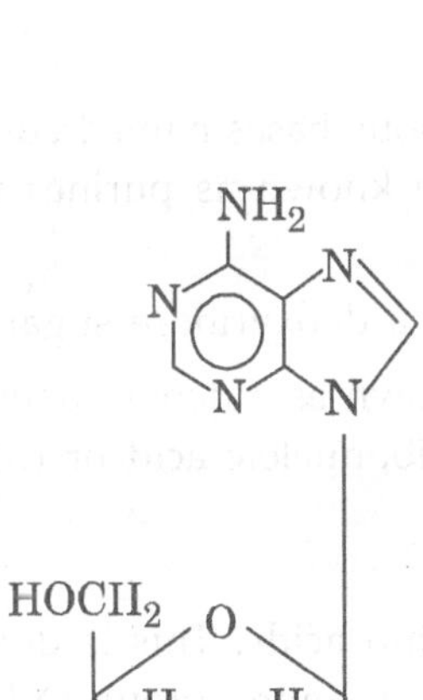

Adenosine

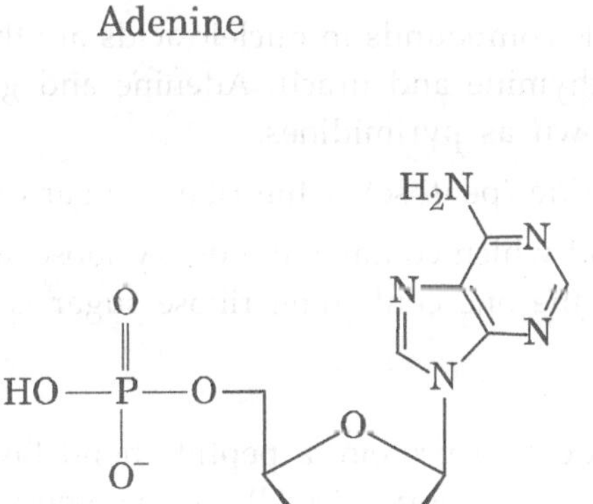

Nucleotide

The nucleic acids DNA and RNA consist of nucleotides only, DNA and RNA function as the genetic material.

1.4.1 Bio Macro Molecules

Biomolecules, i.e., chemical compounds found in living organisms are of two types. The first ones have molecular weights less than one thousand Dalton and are usually referred to as micromolecules or simple biomolecules and the second ones are those that have water. The water insoluble fraction are macromolecules or biomacromolecules.

1.5 PROTEINS

Proteins are polypeptides. They are linear chains of amino acids linked by peptide bonds. The polymer of amino acids make proteins. There are 20 amino acids (alanine, cysteine, proline, tryptophan, lysine etc.) which make up proteins.

The amino acids are essential or non-essential. The essential ones are those that we get from our diet or food and the non-essential ones are those that our body makes.

In living organisms, proteins have different roles to play. The nutrients are transported through the cell membrane with the help of proteins. Organisms that cause infection in the body are fought by proteins. Hormones or Enzymes are also made up of proteins.

1.6 NUCLEIC ACIDS

Nucleic acids are also macromolecules to be found in the acid insoluble fraction. The building block of a nucleic acid is a nucleotide and for a living tissue they comprise the true macromolecular fraction. As stated above, a nucleotide has three components:

1. A heterocyclic compound
2. A monosaccharide
3. A phosphoric acid or a phosphate

The heterocyclic compounds in nucleic acids are the nitrogenous bases named adenine, cytosine, guanine, thymine and uracil. Adenine and guanine are known as purines while the other 3 are known as pyrimidines.

A monosaccharide (pentose) is the ribose sugar or it can be a deoxyribose sugar also.

The nucleic acid which contains the deoxyribose sugar is known as a deoxyribonucleic acid or (DNA) and the one containing ribose sugar is called a ribonucleic acid or (RNA).

1.6.1 Bonding

In a polypeptide or a protein, a peptide bond links the amino acids. This is done by the linking of a carboxyl group (COOH), of one amino acid with an amino group (NH_2) of another amino acid. A water molecule is also released in the process.

Glycosidic bond is the one that links many single monosaccharides into a polysaccharide. This is done by the linking of two carbon atoms of two adjacent monosaccharides. A water molecule is released in this process also.

In case of a nucleic acid, a phosphate links the 3′ carbon of one sugar of one nucleotide to the 5′ carbon of the sugar of the succeeding nucleotide. The bond between the phosphate and the hydroxyl group of sugar is an ester bond.

Below are shown the 3′ and 5′ ends of a ribose sugar:

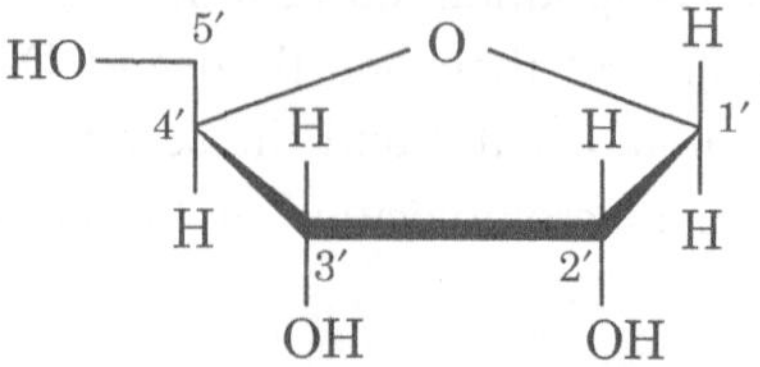

Ribose

The atoms in the ribose ring are numbered as 1′, 2′ etc. merely to distinguish them from the base positions.

A nucleic acid chain, of whatever length has a free 5′ end which may or may not have any attached phosphate groups and a free 3′ end which is most likely to be a free hydroxyl group.

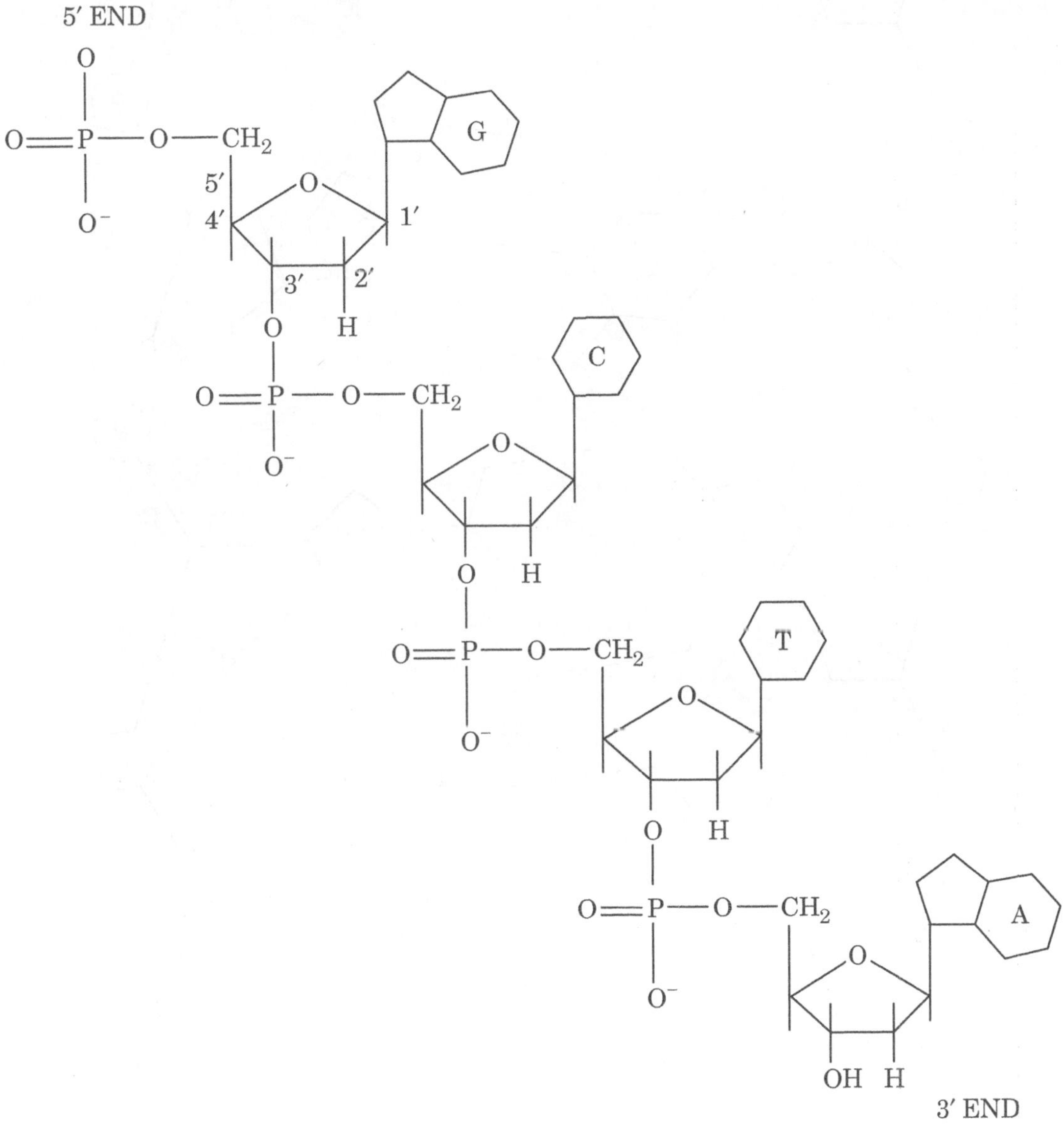

Fig. 1.6: The Nucleic Acid Side Chain

Another picture would look like this:

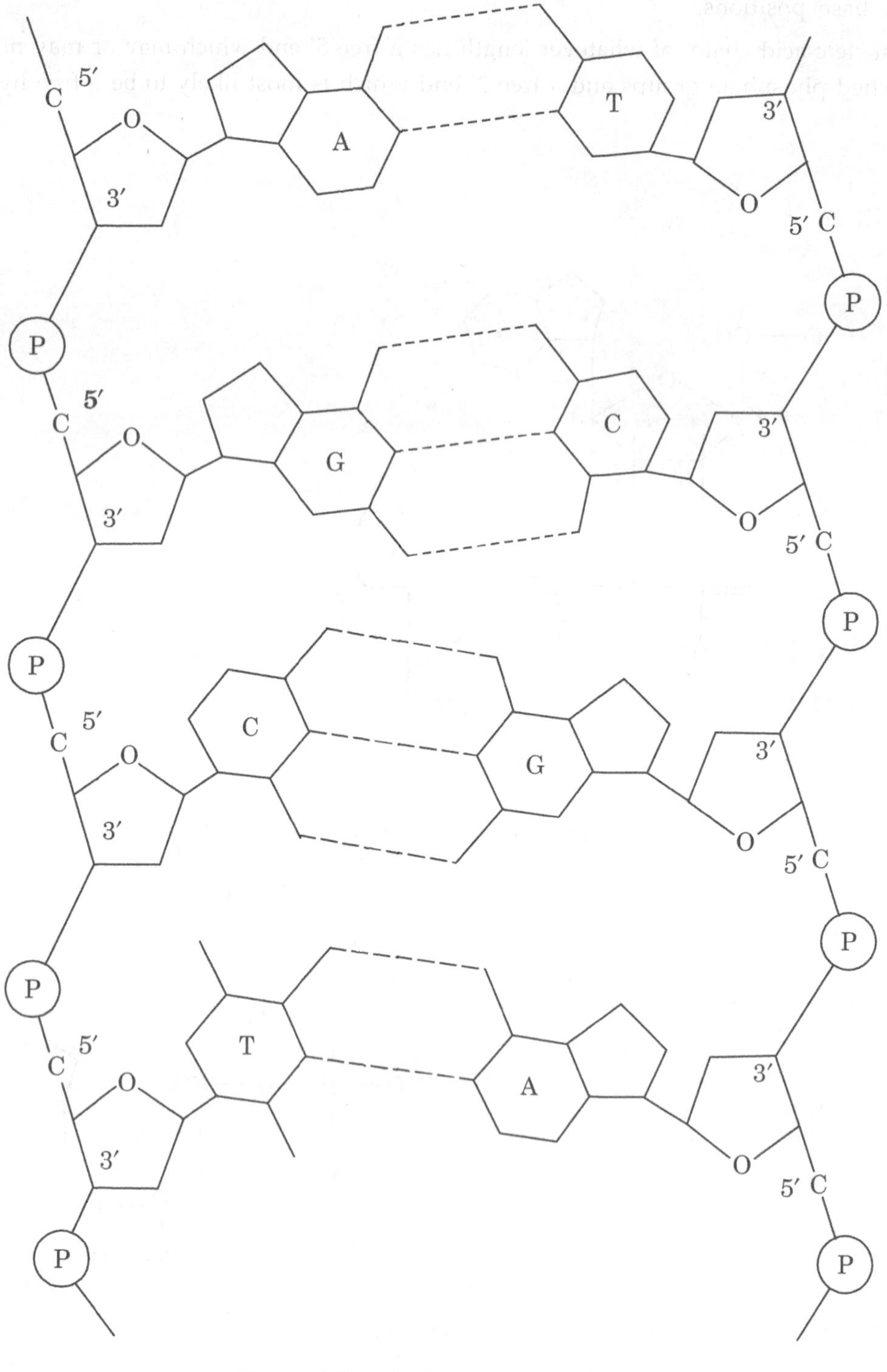

Fig. 1.7: The Nucleic Acid Side Chain

The model shown above is the *Watson and Crick model* which says that DNA exists as a double helix. The two polynucleotide chains are placed antiparallely to each other, i.e., they move in the opposite directions to each other. The backbone is formed by the sugar - phosphate - sugar chain.

The nitrogen bases (adenine, cytocine, thymine and guanine) are projected more or less perpendicular to this backbone but face inside. A and T pair compulsory with each other and G and C with each other. There are 2 hydrogen bonds between A and T and 3 hydrogen bonds between C and G.

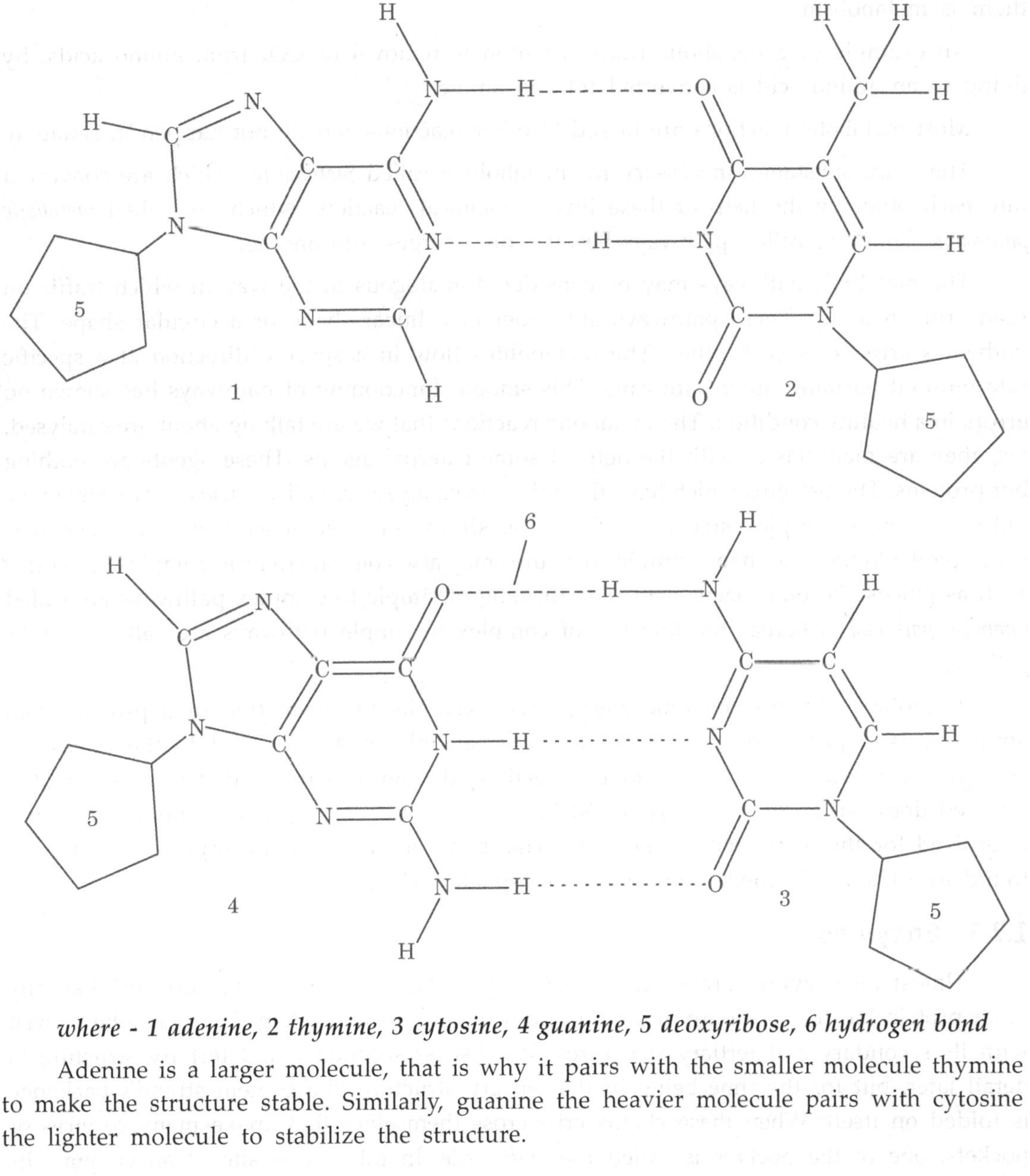

where - 1 adenine, 2 thymine, 3 cytosine, 4 guanine, 5 deoxyribose, 6 hydrogen bond

Adenine is a larger molecule, that is why it pairs with the smaller molecule thymine to make the structure stable. Similarly, guanine the heavier molecule pairs with cytosine the lighter molecule to stabilize the structure.

1.7 METABOLISM

A plant or an animal contain thousands of organic compounds. These compounds or biomolecules are present in certain concentrations (expressed as mols/cells or mols/litre etc). These molecules are constantly being changed into some other biomolecules and also made from some other biomolecules. The molecules that break and the new molecules made are done via the chemical reactions which are taking place everytime in living organisms. These chemical reactions in turn are called metabolic reactions and the term used to describe them is metabolism.

An example of a metabolic transformation is removal of CO_2 from amino acids. By doing so an amino acid is converted into an amine.

Most metabolic reactions are linked to other reactions and do not happen in isolation.

There are substances necessary for metabolism called *metabolites* which are converted into each other by the help of these linked chemical reactions which are called *metabolic pathways*. They are called pathways because one merges into another.

The metabolic pathways may be considered analogous to the way in which traffic on roads run in a city. These pathways are either of a linear shape or a circular shape. The pathways criss-cross each other. The metabolites flow in a specific direction at a specific rate without harming other pathways. This smooth functioning of pathways has shown no errors in a healthy condition. The metabolic reactions that we are talking about are catalysed, i.e., they are made faster with the help of some external agents. These agents are nothing but proteins. The proteins which have the ability to catalyze are called *enzymes*. The metabolic pathways form complex structures from the simple ones such as acetic acid becomes cholesterol. On the other hand, simple structures may also combine to form complex structures such as glucose becomes lactic acid. The forming of simple to complex pathways are called *anabolic pathways* whereas the forming of complex to simple pathways are called *catabolic pathways.*

Anabolic pathways consume energy. For example, the formation of a protein from amino acids requires energy. And conversely catabolic pathways lead to the release of energy, e.g., when glucose is degraded to lactic acid, energy is released. The energy that is released does not go waste but is stored in the form of chemical bonds. This bond energy is utilized for the work that we perform. This bond energy in living organisms is mostly found in a chemical called Adenosine triphosphate (ATP).

1.7.1 Enzymes

Almost all enzymes are proteins. An enzyme has a primary structure just like any other protein has, i.e. made up of amino acid sequence of the protein. It can also be viewed with its secondary and tertiary structures. We discuss secondary and tertiary structure in detail later, but for the time being in the tertiary structure, the protein strand's backbone is folded on itself. When these chains criss-cross themselves they make many cervices or pockets, one of the pockets is called the *active site.* In this active site of an enzyme, its substrate falls and sits there. These enzymes through their active sites catalyze reactions at a very high rate.

1.7.2 Chemical Reactions and Enzymes

To understand how enzymes work, we first have to understand a chemical reaction. Chemical compounds may undergo a physical change or a chemical change. In a physical change the shape is only changed and no making or breaking of bonds is taking place. Also when ice melts into water or water becomes a vapour, there is a change in the state of matter and it is also referred to as the physical change. However, when bonds are broken or new bonds are formed then a chemical reaction is supposed to occur.

For example: $Ba(OH)_2 + H_2SO_4 \rightarrow BaSO_4 + 2H_2O$

The above reaction is an example of an inorganic chemical reaction.

Catalyzed reactions proceed at rates that are higher than uncatalyzed ones. When enzymes catalyze reactions, the rate of processing of a reaction is much higher than uncatalyzed ones. In a multi-step chemical reaction, different steps of a chemical reaction are catalyzed by same or different enzymes, and this is what is referred to as a *metabolic pathway.*

1.7.3 Cofactors

Enzymes are composed of one or more polypeptide chains. (Chains of amino acids of proteins). In other cases, some non-protein constituents called *cofactors* are also attached to the enzymes to make the enzyme catalytically active. Here, the protein portion is called the *apoenzyme* and the non-protein is called the *cofactor* of an enzyme. The catalytic activity may be lost from an enzyme when the co-factor is removed from it. This signifies that they play a substantial role in the catalytic activity of an enzyme.

Unit 2

Introduction to Bioinformatics

2.1 DEFINITION

Bioinformatics in its simplest form can be defined in a single sentence as "It is the intersection of biology and Information Technology". Or we may state it as "It is the science of using Information Technology to understand Biology".

We all know that biology is the study of living things and why do we need information technology to understand biology. The answer is simple, biology if looked at in greater depth has a lot of information, which needs to be analyzed, compared, interpreted in order to understand the functions and physical characteristics of living organisms. And what is that information. That information is nothing but the hereditary material present in the cell (the DNA or the genes). Because this information is vast, technology is required. A human cell consists of 3 billion base pairs in a DNA strand. (A base pair is an A-T or C-G pair). Technology helps because data collection can be done with the help of sophisticated laboratory technology. And data can also be interpreted with the help of technology using sophisticated algorithms.

In bioinformatics a couple of points need to be addressed:

1. How do we know which part of DNA sequence is responsible for which function of living organism.
2. Can we determine the structure of the protein correctly if only its sequence is known.
3. DNA is finally translated into protein. How do we know which part of the DNA translates to which protein.

Bioinformatics is the field which helps us to address the above points. Bioinformatics depends on methods used by statisticians and also techniques used for pattern recognition. We can do a number of experiments for DNA comparison by developing useful algorithms, databases, user interfaces and statistical tools. With the above mentioned tools, the results provided can be of considerable importance.

There is a drawback with bioinformatics and that is, if the person using the tools is not fully aware of what is the meaning of the data that he is using and how the data is generated, he may be lost in the information pool and produce wrong results.

The goal of bioinformatics is to know how living organisms function. In order to do so it uses the fields of statistics, physics, computer science and engineering to capture, manage and present data in the most useful form. Further refinement requires sequence comparison, sequence database searching and sequence analysis and population of databases, and user interface development. In doing so complicated algorithms are designed for the purpose.

The step wise process of bioinformatics involves the following:

1. Abstraction of the properties of biological systems.
2. Making mathematical or physical model out of it.
3. Implementing the algorithms to analyze it.
4. Develop and populate the databases.
5. Use of interfaces such as web tools to access them.

2.2 INFORMATICS AND BIOLOGY

We know that informatics is the term used in connection with organization, representation, distribution, maintenance and manipulation of data in digital form.

The first feature of bioinformatics is the representation, storage and distribution of data. In this feature we cover the design of data formats and databases, development of tools to query those databases and then making of user interfaces to use the tools.

The second feature of bioinformatics is to discover the knowledge that is hidden in the data with the help of the tools. By knowledge we indicate any type of information that surfs up while we try to answer the questions put up by biological scientists.

Bioinformatics needs to find solution to the following problems:

1. To find the function of a gene in a DNA sequence.
2. The protein folds into its native 3D structure, we need to find out how it folds into this 3D structure from a linear chain of amino acids (primary structure).
3. To find out the functioning of a cell, by looking at the various reactions that are taking place inside the cell.

To summarize it all, we again say that the goal of bioinformatics is "To find out how living organisms function and why do they show visible characteristics (i.e., phenotype, e.g., blue eyes) given the genome sequence(DNA sequence)."

2.3 AREAS OF BIOLOGY WHERE BIOINFORMATICS HELPS

There are a number of areas of biology where bioinformatics can be used to solve the individual problems:

1. The entire DNA content present in various chromosomes of a cell is called the genome. The largest area of biology deals in finding out how this genome develops into a living organism.

2. In the genome, are present various genes. Till date scientists have identified a lot of genes which are responsible for various working of organisms, but there remains still a lot of unknown genes, whose functions have yet to be open.
3. We know that the linear chain of amino acids of a protein fold into a 3D structure. But what is the function associated with that protein and the relationship between the structure and the function of the protein needs to be found out further since the proper foretelling methods are not available.
4. The biological databases, e.g., GenBank, holding biological data is filled with vast amount of data. This data needs to be compiled properly and offerred to the user properly in an intelligent format. This huge data management is a complex task and involves the science of Human Computer Interaction for the demonstration of data.
5. Managing this huge data is another complex task because the data is interlinked. The scientists need to be very careful in following these links and not getting lost in the information pool.
6. Genes do not exist independently but co-exist with other genes and interact with each other to form biochemical pathways. These pathways meet other pathways to perform some function. For scientists, it is easy to identify the functioning of a single gene, but when the gene work together, the task becomes highly complex.
7. There are a lot of external agents that interact with the genes. These external agents form complex biochemical reactions with the genes. These reactions have to be studied and analyzed to again find out the functioning of the cell.

2.4 HOW COMPUTERS HELP IN BIOLOGY?

1. If we analyze a single gene then it leads to one biochemical pathway or protein analysis. But we all know that single genes do not perform on their own rather they interact with other genes or we can say one protein or biochemical pathway interacts with other proteins or other biochemical pathways. If multiple genes are put for experimentation, then a lot of people need to work and their ideas need to be communicated to other research groups. To do all this, the internet is used to collect the ideas from various research communities, put them at one place and then distribute it to whoever requires it.
2. Another area where computers help is the area of automated sequencers. In automated sequencers the sequences of human clone are collected and compared with the sequences of other organisms of some notable genes whose tasks have already been identified. These notable genes show some uniqueness which are related to the phenotype of the organism. By phenotype we mean showing some visible features which can be easily spotted out. For example, the colour of hair is blonde or black. The shape of nose etc.
3. In earlier days chemical mapping techniques were used for finding out the maps of the genomes of living organisms. These days those techniques are becoming outdated and

new techniques such as image processing are in picture. These techniques may take several years to give a detailed genome map of human beings.

4. X-Ray crystallography is another area where computers are used. The reflections of proteins are available in minutes instead of hours or days or months. There are methods of automated analysis which find out the protein crystallization structures in days or weeks time.

   The technique of X-ray crystallography is described below.

   In 1912, M. Von Lane discovered that X-rays could be diffracted in expected patterns by crystals of a regular array of molecules. In 1950, D. Hodgkin found a method to determine the structure of proteins by observing the diffracted beam of X-rays when protein molecules were crystallized and diffracted. X-ray crystallography has successfully obtained more than 15,000 structures of proteins at very high resolution.

   The procedure is as follows:

   At first, the protein crystals are grown for their structure determination. Once the crystals are made, they are filled in a capillary tube and X-rays are passed through them, which diffracts the protein crystals. The diffracted patterns hit the screen and are captured by a detector. These patterns are then fed into the computer for analysis. The analysis of the diffracted patterns is done by means of complex software tools and reverse Fourier transformations.

   The analysis leads to the three dimensional structure of proteins.

   The Protein Data Bank (PDB) is a database which contains the entries of protein structures which are obtained from X-ray crystallography.

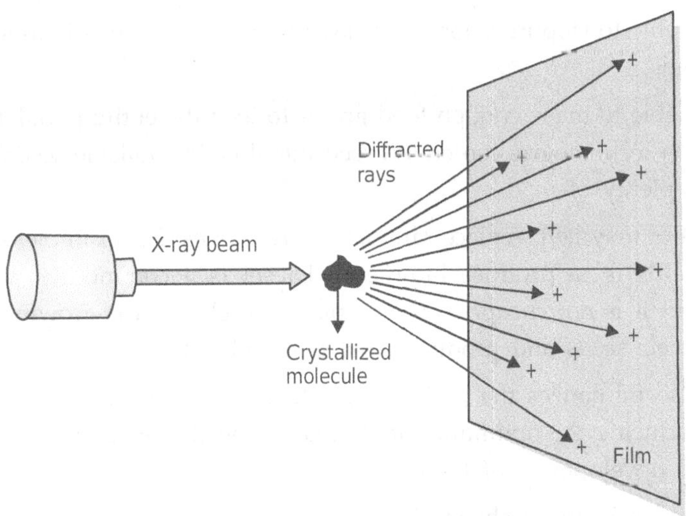

Fig. 2.1: X–ray Crystallography

5. You must have heard of parallel computing. In this, the problem is broken down into manageable components and each component is solved by a different processor simultaneously.

The experiments involving DNA microarrays requires several genes to be experimentally tested. Thousands of gene expression experiments are carried out in parallel on a tiny chip. Computers are required to collect, process and analyze the data simultaneously so that the results of the experiments can be inferred.

Let us look at the reasons which caused the start of automated work in molecular biology from manual work,

1. The beginning of storage of raw biological data into public databases made it possible for researchers to share their work. The scientists could now save a lot of time by sharing common points of interest and not re-inventing the wheel. The area where research had already taken place was now available to everybody and therefore they need not spend any more time working on it.
2. The second area which brought the whole community together was the world wide web. This made it possible for the entire research community to share information and distribute information.
3. Another trendsetter was the computerization of experimental molecular biology. A lot of experiments were carried out by human hands in earlier days. These days computerized experiments make the job much easier and faster. The research in molecular biology has taken an entirely different shape.

2.5 BIOINFORMATICS SOLVES THESE PROBLEMS

1. Bioinformatics helps in curing diseases. If we know how genes help in the working of living organisms then we can find the related diseases and possibly find their cure too.
2. We may be able to stop infection. If we know what causes the diseases, we may be able to stop them.
3. We may be able to make enough food grains to feed the entire population. A lot of new hybridization technologies can cross breed the already available seeds and new verities can supply plenty of food.
4. Bioinformatics may help replace rare natural resources. For example, an upcoming area where it helps is to produce bio-diesel. Diesel is a commodity which is naturally available, but it is not cheap. To help replace diesel, India is growing plants to extract oil out of their seeds and produce what is called bio-diesel.
5. This fuel also minimizes the pollution produced by burning diesel. So, bioinformatics helps in producing the optimum quality plants and the research that is required in the growth and development of these plants.
6. Other drugs, agricultural chemicals, plastics and petroleum derivatives are developed by many companies to look after the growing needs of consumers and to produce variety among these products and to select the optimum one. All this needs a lot of research and development to be carried out by bioinformatics scientists.

Let us draw a flow chart to re-summarize the goal of bioinformatics.

To identify each gene

↓

Match the gene with the protein equivalent

↓

Find the function and structure of each protein

↓

Find out how living organisms work.

2.6 PROTEIN SEQUENCING

DNA, RNA and protein store an organism's genetic and functional information. These are all made up of linear chains of smaller molecules called macromolecules.

The macromolecules are derived from a fixed number of chemicals:

1. DNA is made of four deoxyribonucleotides (adenine, thymine, cytosine and guanine).
2. RNA is made of four ribonucleotides (adenine, cytosine, guanine and uracil).
3. Proteins are made from 20 amino acids.

These macromolecules can be expressed as a sequence of symbols because they are chains of linear components. These sequences can then be compared to set up similarity among the molecules which may be linked by form (phenotype) or function. This is one of the oldest method used to find similarity amongst living organisms by the scientists.

The earlier methods were concerned with identification and quantification of amino acids. Insulin was the first peptide hormone that was sequenced in about 1945. By 1950, an enzyme by the name of ribonuclease had its sequence ready. In 1965, round about 20 proteins had their sequences completed. By 1980, the number of sequences grew to 1500, and the number of protein sequences available today are in the order of millions.

In 1960s and 1970s, the first nucleic acid sequencing of the RNA was done. This was a tRNA or a transfer RNA which has round about 95 nucleotides. DNA has longer sequences of nucleotides ranging from 55 million to 250 million base pairs. In early days sequence comparison was done by the human eye, but these days two well developed programs are available by the name of BLAST and fsBLAST, for comparison of amino acid sequence and DNA sequences respectively. These days a molecular biologist can take any DNA sequence and compare it to publicly available DNA sequence pool for finding similarities.

Let us take an example to demonstrate the use of the BLAST program. This will help us uncover how diseases are related to genes.

The *Drosophilla melanogaster* or more commonly known as the fruit fly have a gene called the eyeless gene. If we remove this gene from the fruit flies with the help of molecular biology techniques and then cross breed them. The fruit flies will develop without the eyes. Therefore, in *drosophila melanogaster*, this gene is responsible for the development of eyes. Another similar gene is present in the humans. If we remove this gene from the human genome then humans develop a disease called aniridia, which is that the eyes develop without the iris.

Scientists further experimented that if this aniridia gene was added into an eyeless fruit fly, the further offspring developed with the eyes. This shows that there is similarity between the eyeless gene of the fruit flies and the aniridia gene of the humans. Because in humans the eyes are more complicated, it does not lead to eyes missing completely but develops with a disorder and in fruit flies the similar gene removes the eyes completely. This proves they have a similar function.

This experiment leads to the conclusion that organisms are similar to each other in functionality or phenotype (physical characteristics). And it is nothing but the DNA sequence or the genome sequence which is responsible for all this since it carries the genetic material. We will compare the two sequences of these organisms and find out where does the similarity lie.

About 40 years ago, scientists used to compare sequences with the hand, character by character. This was a very boring and time consuming process and used to strain the eyes. BLAST and fsBLAST are designed to compare the sequences digitally rather than by hand. Pairwise comparison of sequences is still the most widely used and trusted mechanism to find the similarities between the genes. The sequences are matched within seconds, rather than taking the scientists hours to do the same amount of work.

2.6.1 Labels

The DNA and protein sequences can be given labels, which are nothing but the sequences of the alphabets which determine their functionality. These sequences are similar to the barcode. The sequences of alphabets are made of the nucleotides or the amino acids, (e.g., ACGTATCC) in case of DNA sequences and (HLKTMLQRV) in case of amino acids of proteins.

By looking at these alphabetic sequences, comparison can be made and useful information can be derived. If not a 100% match is there, still important findings are made and conclusion can be drawn by looking at these patterns. Here when gene differ or show similarity, we draw meaningful conclusions and this is the reason why these organisms show differences or similarities in function or phenotypes (physical characteristics).

Simple labels cannot strictly be applied in finding biological data, since they may have undergone development, means the data may have changed slightly or considerably. So, we need to be careful, that related data may be considerable findings for us. This all happens because of mutations, so the BLAST program with 100 % matches or partial matches may be of use to us.

2.6.2 Comparison of Eyeless Gene and Aniridia with BLAST

Query: 22 HSGVNQLGGVFVGGRPLPDSTRQKIVELAHSGARPCDISRILQVSNGCV
SKILGR YYETG 116
HSGVNQLGGVFV
GRPLPDSTRQKIVELAHSGARPCDISRILQVSNGCV SKILGRYYETG
Sbjct: 3 HSGVNQLGGVFVNGRPLPDSTRQKIVELAHSGARPCDISRILQVSNGCVS
KILGRYYETG 63

Query: 117 SIRPRAIGGSKPRVATAEVVSKISQYKRECPSIFAWEIRDRLLQENVCT
NDNIPSVSSIN 176
SIRPRAIGGSKPRVAT EVVSKI+QYKRECPSIFAWEIRDRLLE
VCTNDN IPSVSSIN
Sbjct: 64 SIRPRAIGGSKPRVATPEVVSKIAQYKRECPSIFAWEIRDRLLSEGVCTN
DNIPSVSSIN 122

Query: 177 RVLRNLAAQKEQ 188
RVLRNLA++K+Q
Sbjct: 123 RVLRNLASEKQQ 133

Query: 189 TEDDQARLILKRKLQRNRTSFTNDQIDSLEKEFERTHYPDVFARERLAG
KIGLPEARIQV 455
+++ Q RL LKRKLQRNRTSFT +QI++LEKEFERTHYPDVFARERLA KI
LPEARIQV
Sbjct: 154 SDEAQMRLQLKRKLQRNRTSFTQEQIEALEKEFERTHYPDVFARERLAA
KIDLPEARIQV 201

Query: 456 WFSNRRAKWRREEKLRNQRR 488
WFSNRRAKWRREEKLRNQRR
Sbjct: 202 WFSNRRAKWRREEKLRNQRR 232

A letter shows an exact match, a space shows no match and a + shows that they have similar physiochemical properties.

As we can see from the above, there are no 100% matches but there are still considerable matches and these are the matches that result in the similarity in the function of eye development of both the organisms.

2.6.2.1 Working of BLAST

BLAST works as follows. It finds local regions that match in pairs and are not exactly the same. In a sequence, it matches a character and then extends the match in all directions until the overall score is small enough. Using this method, distantly related matches can be found rather than the ones which are matched perfectly. And as can be seen from the experiment discussed above that results that are significant from biological point of view can be fetched from partial matches rather than exact matches.

Unit 3

MOLECULAR BIOLOGY DATA

3.1 INTRODUCTION

Public Biological Databases have been developed to hold the enormous molecular biology data. These databases are established for the scientists across the world to research and exchange their software and data. They are connected together with the help of the WWW.

3.2 PUBMED

PubMed is one of the most widely known and useful web resource for the scientists. It has journals for the topics of molecular biology, biochemistry and genetics. Search in PubMed is made by Keyword searches and Boolean operators AND, OR or NOT are used to form the queries. Keywords from the Medical Subject Heading (MeSH) can be searched in PubMed.

3.3 MOLECULAR BIOLOGY DATABASES

The names given to the genes are derived from developmental biology studies or from cellular biological studies, that describe the function of a gene at a cellular level. They may also come from studies relating to proteins.

The names given to proteins are normally not of the gene from whose breakdown these proteins have been formed. So, we need to give a common name to the gene, the gene product (protein sequence, protein structure, DNA sequence) biochemical process.

A number of factors can be related to a gene, its source organism, its chromosomal location, the regulatory proteins. The proteins can further be identified as structural protein, an enzyme, or a functional RNA. They can also be identified by the product it produces, by its structure and lastly by the knowledge of the tissues where expression occurs.

This is the sort of information that is linked to a gene.

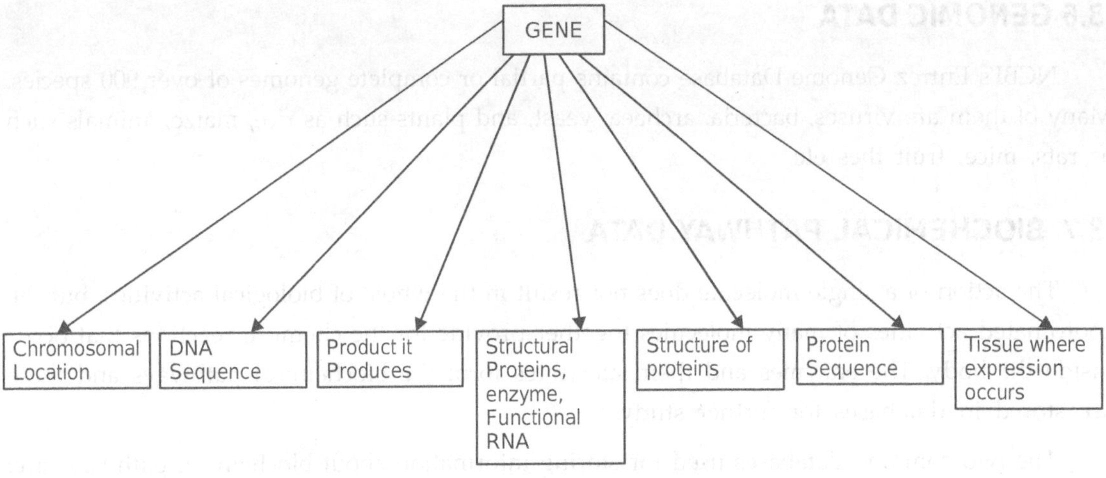

Fig. 3.1: Information of a Gene

The information about a gene is stored in the database, in the following fashion. First an entry is made into the database with a serial number and then stored it in a relational database that follows the links between the serial number and the names of the gene, the gene product or any other information about the gene. The gene product can be represented in different forms such as protein sequence, the DNA sequence and protein structure. All these data types are maintained separately though they represent the same gene product.

3.4 PROTEIN STRUCTURE DATA

Protein Data Bank (PDB) was the first public molecular biology Database that held the data for the X-ray crystal structures of protein molecules. In 1971, the PDB was established at the Brookharem National Laboratory, to store protein structure data. In 1976, the number of entries were 76, in 1988 they were 400. Till date there are more than 25,000 entries. The drawback of PDB is that it suffers from redundancy. A lot of proteins have many crystal structures taken for them which now exist in the database.

3.5 DNA, RNA AND PROTEIN SEQUENCE DATA

Sequence data is normally of one type either DNA, RNA or protein. These data are collected mainly in Europe, Japan and the United States of America. Gene Sequence Database (GSDB) was the first DNA sequence Database which was established in 1979. GSDB is now supported by GenBank, but National Centre for Genomic Research still provides information for GSDB.

Other centers which provide data to the GenBank are the European Molecular Biology Laboratory, the National Institute of Health and the DNA database of Japan. The NCBI or the National Center for Biotechnology Information has developed a standard relational database format for sequence data.

3.6 GENOMIC DATA

NCBI's Entrez Genome Database contains partial or complete genomes of over 900 species. Many of them are viruses, bacteria, archaea, yeast, and plants such as rice, maize, animals such as rats, mice, fruit flies etc.

3.7 BIOCHEMICAL PATHWAY DATA

The action of a single molecule does not result in the whole of biological activities, but the coordinated activities of many molecules together provide for the chemical reactions that occur inside the body. The enzymes and their substrates form the biochemical pathways and these are stored in databases for further study.

The two common databases used for storing information about biochemical pathways are:

1. WIT
2. KEGG

WIT stands for "What Is There". It has metabolic pathways of organisms that have their genome sequences entirely sequenced. KEGG stands for Kyoto Encyclopedia of Genes and Genomes and it has similar information. It further has links to structure, sequence and genetic databases.

3.8 GENE EXPRESSION DATA

DNA microarrays (Gene Chips) are small laboratories for the study of gene expression. Each chip contains an array of search molecules that can bind specific pieces of DNA or mRNA to themselves. These search molecules are fluorescent and therefore when bound to DNA or mRNA stand out. Further because of this, the level of gene expression has been able to be calculated in quantitative terms.

This technology was developed in late 1990s and this gives rise to the availability of data for gene expression. The two most popular databases for this are "National Human Genome Research Initiative's Microarray Project Site" and "Stanford Genome Resources Site". Such databases are queried by giving the name of the gene in the query or its functional description in the query.

The table given in section 3.9 shows all the databases that are related to a gene, gene product or biochemical pathway. We will further describe them briefly.

NCBI and EMBnet are Information sources and the rest are the databases. We will first give the table, then talk about the information sources and finally talk about the databases.

3.9 MAJOR BIOLOGICAL DATA AND INFORMATION SOURCES

S. No.	Database Related to	Name of Database
1	Biomedical Literature	PubMed
2	Nucleic Acid Sequences	GenBank EMBL_SRS DDBJ GSDB
3	Genome Sequences	Entrez Genome TIGR Data base (TDB) UniGene
4	Protein Sequences	GenBank SWISS-PROT PIR MIPS
5	Protein Structure	PDB
6	Biochemical Pathway	PathDB KEGG WIT
7	Composite Protein Sequence Database	NRDB OWL MIPSX SP + TrEMBL
8	Secondary DataBases	PROSITE Profiles PRINTS Pfam Blocks IDENTIFY
9	Structure Classification Databases	SCOP CATH PDBSUM

3.9.1 EMBnet (European Molecular Biology Network)

In 1988, a network was recognized to tie European Laboratories that were working in the area of bioinformatics and biocomputing. The network's main focus was to provide information, services and training to users spread across the different European Laboratories via chosen nodes which were operating in their own countries. This effort made a central place for holding data of diverse biological databases.

EMBnet controls 34 nodes in all, out of which 20 are of different countries. These are maintained by the Governments of these countries. The services supplied by these nodes include sequence analysis, protein modeling and genetic mapping etc. Apart from this, the complete database includes, software having user support and it also provides training to those who are involved in research and development.

Another 8 nodes are known as "Specialist Sites". These are the academic, industrial or research centers which hold information about specific areas of bioinformatics. Example, the EBI site of UK has EMBL nucleotide database.

List of Specialist Sites

MIPS	Germany
ICGEB	Italy
Pharmacia Upjohn	Sweden
F.Hoffman-La Roche	Switzerland
EBI	UK
HGMP-RC	UK
Sanger	UK
UCL	UK

Another 6 sites are known as "Associate" nodes. These are the supporting sites which provide information to the European networks, but are from non-European countries.

List of Associate Nodes

IBBM	Argentina
ANGIS	Australia
CBI	China
CIGB	Cuba
CDFD	India
SNABI	South Africa

3.9.1.1 EBI

EBI is commonly known for the European Bioinformatics Institute which was recognized in 1994 and is a support branch to the European Molecular Biology Laboratory (EMBL).

The tasks of the EBI include:

1. To make and give out the Nucleotide Sequence Database which is part of the EMBL group, Europe's source of Nucleotide Data Sequences.

The Japanese counterpart source of nucleotide sequences is known as the DDBJ and the United States's corresponding project is known as GenBank which belongs to the NCBI group. Both the DDBJ and GenBank work in partnership with EMBL Nucleotide sequence Database. The EBI ensures that all the latest developments (New and Updated sequences) are passed on to these two groups everyday.

2. The EBI's second responsibility is to work hand in hand with the Swiss counterpart which is known as the Swiss Institute of Bioinformatics. It therefore maintains and distributes SWISS-PROT entries which is the sequence of protein chains.

3.9.1.2 MIPS

Another Specialist site of the EMBnet is MIPS, which is commonly known as the Martinsried Institute for Protein Sequences (MIPS). MIPS also works in partnership with an international body for holding protein sequences which is known as PIR-International protein sequence. There the MIPS is representing the European continent. The tasks of MIPS are again to build up, revise, and give out the protein sequences within European Nations.

The Institute also caters to the needs of people wanting to know more about related protein families. Also it allows searches to be made to homologues and allows the alignments to be inspected.

3.9.1.3 UCL

UCL is another of the specialist node of the EMBnet. This node is meant for looking after the protein sequences and their structure analysis. This node was recognized in 1988 and its main focus is to deal with 3 protein sequence related databases, namely, PDBSum database, PRINTS and CATH. The PDBSum database takes care of summaries and structural analysis of the PDB data files, the PRINTS database is mainly responsible for protein fingerprints and finally CATH is a protein structure classification resource.

The node also provides a number of software tools, which can be downloaded from its site. These include software to:

1. plot graphic diagrams of protein-ligand relations.
2. compute hydrogen and non-bonded interactions.
3. study protein structural motifs.
4. confirm the stereochemical value of protein structures.
5. envision structure of proteins.

3.9.2 NCBI (National Centre for Biotechnology Information)

This is the American resource equivalent to the EMBnet. It was established in 1988 under the National Library of Medicine. NCBI is mainly focused on how to expose the hidden knowledge concerning health and disease by using information technology to take out that knowledge from molecular and hereditary methods.

The tasks of NCBI are listed as follows:

1. Formation of computerized systems for piling and studying biological information.
2. Making of sophisticated techniques to process information.
3. Helping users to access the databases and software in an easy manner.
4. Organize the hard work done in accumulating information worldwide.

The main objective of NCBI is to look after GenBank which is the American resource for maintaining DNA sequence database. The records of the DNA sequences are made from scientific literature and the information gathered is distributed to other counterparts, namely, EMBL and DDBJ.

3.9.2.1 Entrez

The NCBI also provides the Entrez facility. The purpose of Entrez is to form an interface where access to multiple databases is provided. The platform gives admission to multiple resources where the same data can be visualized in different ways and hence better understood.

The databases of NCBI are incorporated together and thus Entrez allows retrieval of molecular biology data from these databases.

These include:

1. Provides the right to see DNA sequences from Databases like GenBank, EMBL and DDBJ.
2. Provides the right to see Protein sequences from databases like SWISS-PROT, PIR, PRF, SEQDB and PDB. Also translated protein sequences can be seen from the DNA sequences.
3. Provides the right to see Genome and Chromosome Mapping data.
4. Provides the right to see 3D protein structures from databases like PDB.
5. Provides the right to look at the bibliographic information from databases like PubMed.

3.9.3 DNA Sequence Databases

As we have already mentioned, 3 major databases for DNA sequences exist and they are GenBank, DDBJ, and EMBL. We will discuss each one of them separately.

3.9.3.1 EMBL (European Molecular Biology Laboratory)

EMBL is the nucleotide sequence database of the European Bioinformatics.

Institute (EBI): The database is preserved with the help of information pouring in from the other 2 databases namely GenBank and DDBJ. The database mainly collects information from all over the world and exchanges the newly built and revised versions of sequences on a daily basis between the other two databases.

The EMBL provides a system to get back the information known as the SRS or the Sequence Retrieval System. This system provides and entry point to surf the molecular biology databases. The EMBnet has various nodes which all have different formats to allow usage of their databases. The SRS accesses all these databases through a standardized interface which hides the underlying details of each node's complexities. The SRS makes queries across a range of databases without needing to know the details of the data structures used and query languages.

3.9.3.2 DDBJ (DNA Data Base of Japan)

This database started in 1986 in partnership with the other 2 major DNA databases, namely GenBank and EMBL. The National Institute of Genetics is mainly responsible for developing the DDBJ. It also looks after the daily changes and passes on the information to the other databases. It takes inputs from all over the world via a web based data input tool.

3.9.3.3 GenBank

The largest and the most up to date database among all the 3 major databases for DNA sequences is the GenBank. This is looked after by the NCBI. The NCBI collects sequence data of every organism from every source. It keeps the DNA sequences in different forms so that more inferences can be drawn such as:

1. mRNA DNA
2. cDNA Clones
3. Expressed Sequence Tags (ESTs)
4. Genome Sequence Data
5. Sequence Polymorphism Data

Let us look at these types of DNA sequences in detail.

1. **mRNA:** Messenger RNA, the product of transcription of genomic DNA. The sequences may be partial or complete. mRNA may be reduced by the cell to remove introns(in Eukaryotes). The mRNA does not cover the complete coding sequence of a gene.
2. **cDNA:** A DNA sequence which is unnaturally made by doing the process of reverse transcription on an mRNA strand. The cDNA or complementary DNA gives a rough approximation of the coding regions of genomic DNA from which the matching mRNA was formed.
3. **Genomic DNA:** It is the DNA strand that contains both the coding and the non-coding regions of the DNA strand. It contains both the introns (non-coding regions) and the exons (coding regions). The genomic DNA is a result of many sequencing processes that have been applied on a single genome stretch and is therefore a complete and dependable piece, from which useful deductions can be made with a good estimate.
4. **ESTs:** It is a short form for Expressed Sequence Tag. These are short cDNA sequences prepared from mRNAs which are extracted from the cells while the cells are developing using molecular biology tools and processes. The ESTs mainly help in identifying a particular gene rather than showing the entire coding sequence of a gene.
5. **GSS:** It is commonly known for Genome Survey Sequence. It is a result of a single pass of sequencing experiments from a genome. It may contain a high percentage of sequencing errors. The difference between GSS and genomic DNA is that the GSS is a result of a single pass sequencing experiment whereas the genomic DNA has resulted after multiple

passes on sequencing experiment over the genome strand. That is why Genomic DNA is more accurate. The GSS is mainly useful if you are searching for a new gene rather than looking for other processed and well identified genes.

A GenBank Sample Record.

LOCUS SCU11234 2012 bp DNA PLN 18-JUL-1996

DEFINITION Saccharomyces cerevisiae TCP2-beta gene, cds is partial

ACCESSION U12567

VERSION U12567.1

GI:3241123

KEYWORDS .

SOURCE Saccharomyces cerevisiae (baker's yeast)

ORGANISM Saccharomyces cerevisiae

Eukaryota; Fungi;

REFERENCE 1 (bases 1 to 2012)

AUTHORS A. James, G. Geetha , N. Swamy

TITLE Cloning and sequence of REV2, the function of this gene is to help in repairing of DNA in Saccharomyces cerevisiae

JOURNAL Yeast 2 (6), 1101-1104 (1991)

PUBMED 3563366

REFERENCE 1 (bases 1 to 2012)

AUTHORS Harry N. Golden G. Lal R.

TITLE Selection of repair sites in yeast

JOURNAL Genes Dev. 2 (1), 111-134 (1991)

PUBMED 3356363

REFERENCE 2 (bases 1 to 2012)

AUTHORS Murli L.

TITLE Direct Submission

JOURNAL Submitted (13-JAN-1991) A. James in USA

FEATURES Location/Qualifiers

source 1..2012

/organism="Saccharomyces cerevisiae"

/db_xref="taxon:1242"

/chromosome="IX"

/map="2"

```
CDS      <1..302
          /codon_start=4
          /product="TCP2-beta"
          /protein_id="PPP34522.1"
          /db_xref="GI:324524"
          /translation="MQLPHDGFMPPRASKARTILV
...
DMRLOPQDRSYNIGKARLTV"

gene     123..1021
          /gene="APM1"
CDS      123..1021
          /gene="APM1"
          /note="plasma membrane"
          /codon_start=1
          /function="required for repair of S. cerevisiae"
          /product="Ala.1"
          /protein_id="PPP34522.1"
         /db_xref="GI:324524"
          /translation="PIGDPQNMSSDEASRDMEQOP
....
HRALMPNSMFVDCWPQO"

gene     complement(1779..1999)
          /gene="REV1"
CDS      complement(1779..1999)
         /gene="REV1"
         /codon_start=1
          /product="Rev1p"
          /protein_id="PPP34522.1"
          /db_xref="GI:324524"
         /translation="PIGMNOPQRSDEFGILKVFNMS
```

```
...
KCQRPILVFGHKS"
ORIGIN
     1  acgcttacgt gacctagact atgcctgact attccgttaa agtctggata atgtctgatc
    61  aggtcttgga atgcttgact gctacagtag accgtaggct acgtcgtcga acgtcgtcag
   121  agtacgtttt accgttcgtt acgtcgttca acgtccgtaa acgttcgatt acttgcttac
   ...

  1921  tctgctacag tcagtacgat cgtcagtacg atacagtatc atcgatcata ccattcagat
  1981  acagtacgta cgatataata cacgtacgt caaatcgttt acagtttcag

/
```

Explanation of above fields:

Locus	The LOCUS field has different data elements which are locus name, sequence length, type of molecule, the division of GenBank, and date of modification. The description of all elements is given below.	
	Locus Name	The locus name in this example is SCU11234. The locus name tries to group the entries with similar sequences: the first three characters tell about the organism.
	Sequence Length	Number of nucleotide base pairs (or amino acid residues) in the sequence. Here the sequence length is 2012.
	Molecule Type	The type of molecule that was sequenced. Here the molecule type is DNA.
	GenBank Division	A record is a member of the GenBank division and the division is shown with a three letter abbreviation. Here the GenBank division is PLN. There are 18 divisions in the GenBank database. Some of them are: PLN - plant, fungal, and algal sequences BCT - bacterial sequences VRL- viral sequences PHG - bacteriophage sequences SYN - synthetic sequences
	Modification Date	The date in the LOCUS field is the date of last modification. In this example the last modified date is 18-JUL-1996.

Field	Sub-field	Description
Definition		Describes the sequence; includes information such as name of the source organism, the name of it's gene or the name of it's protein, or some information regarding the function of the sequences (if the sequence is noncoding). If the sequence has a coding region (CDS), description is followed by the word "cds".
Accession		The unique identifier for a sequence record. An accession number is usually a combination of a letter(s) and numbers, normally a single letter is followed by five digits (e.g., U12567).
Version		A nucleotide sequence identification number tells about a single, precise sequence in the GenBank database. In this example it is given by "U12567.1".
GI		"GenInfo Identifier" sequence identification number, here it is meant for the nucleotide sequence. If there is a change in the sequence, a new GI number will be generated.
Keywords		Word or phrase that describes the sequence.
Source		Abbreviated form of the organism name. Here it is "Saccharomyces cerevisiae (baker's yeast)"
	Organism	The formal scientific name for the source organism. Here it is "Saccharomyces cerevisiae"
Reference		The Publications made by the sequence's Authors:
	Authors	List of Authors.
	Title	Title of the published work.
	Pubmed	PubMed Identifier (PMID). References that include PubMed IDs give the number of the PubMed records that match up with this sequence record.
	Direct Submission	The submitter's contact details.
Features		Information about genes and gene products, as well as regions which show some significance in the sequence. The regions include code for protein sequences.
	Source	Mandatory feature in each record that gives the sequence's length, the organism's scientific name, and Taxon ID number.
	Taxon	A unique identification number for the taxon of the source organism. A taxonomy ID number is assigned to each taxon (species, genus, family, etc.) in the NCBI Taxonomy Database
	CDS	Coding sequence; region of nucleotides that corresponds with the sequence of amino acids in a protein (location includes start and stop codons). The feature of the CDS also has an amino acid translation.

	1..302	Base span of the biological feature, is a CDS feature. In this example it is: Eg: <1..302> The feature extends from base 1 through base 302 in the example above, and is partial on the 5' end
	protein_id	A protein sequence identification number, which is equivalent as the Version number of a nucleotide sequence. Protein IDs are made up of three letters followed by five digits, a dot, and a version number. If the sequence changes (even a single amino acid), the version number is increased, but the accession portion remains the same (e.g., PPP34522.1 will change to PPP34522.2).
	GI	"GenInfo Identifier" is the identification number of the sequence, in this case, for the protein translation.
	Translation	The amino acid translation which are equivalent to the nucleotide coding sequence (CDS). In many cases, the translations are conceptual.
	Gene	A region of biological interest identified as a gene and for which a name is given. The gene feature's base span is reliant on the furthest 5' and 3' features.
	Complement	This tells that the feature is placed on the complementary strand.
	Other Features	These include other records that illustrate a number of biological features including a graphic format for each sequence record that visually shows the annotated features.
Origin	The sequence data.	

3.9.4 Genome Sequence Database

3.9.4.1 GSDB (Genome Sequence Database)

This database contains a complete set of DNA sequences to carry out genome sequencing experiments. This database is maintained by National Center for Genome Resources. The database is mainly responsible for accumulating, revising and passing on the DNA sequences to other databases. The producers of sequences can submit their data to this database online. The format of GSDB entries is similar to that of the GenBank entries. Only difference is that the GSDB entries carry a GSDBID Keyword which is basically an accession number of the GSDB entry.

3.9.4.2 TIGR (The Institute for Genomic Research) Data Base (TDB)

This database contains a compilation of different types of data such as protein sequences, DNA sequences, Gene expressions, data related to cell's activities, information about protein families and taxonomic text for animals, plants and microorganisms. This database mainly carries references to many genome sequencing projects of various organisms such as (A. fulgidus, B. burgdorferi, H. influenzae, H. pyrori, H. janneschii and M. genitalium). The TIGR database also has information about parasites such as (T. brucei and P. faliciparum). It also has the database of organisms such as humans, mouse and plants like rice etc. The TIGR database also keeps information about gene index projects and mainly concentrates on the Human Gene Index (HGI) project. The HGI project was developed to find out all the possible genes that exist in the human body. It stores information from the expressions of the genes to the function of the genes and the evolutionary patterns that they show to other genes. It has more than 100,000 ESTs from over 300 cDNA libraries. The data was taken from dbEST and put together with non-redundant Human Transcript (HT) information to generate contigs or Tentative Human Consensus (THC) sequences. The TIGR used a special assembler to carry out the assembling process of the above mentioned information.

3.9.4.3 UniGene

UniGene database shows information about the genes that is actually expressed. From the GenBank database the sequences related to a particular gene help to get the non-redundant gene oriented clusters which further help in producing the transcript maps. The genes of many organisms are taken and each cluster corresponds to a particular gene of interest. It has further information, such as the tissue from where the gene is extracted or where the gene expresses itself, its map, location etc. The database contains a mixture of well distinguished genes and many new ESTs of undistinguished genes.

3.9.5 Protein Sequence Databases

3.9.5.1 SWISS-PROT

This is a database for storing information about the protein sequences maintained by EMBL at Geneva. It was started in 1986 by the department of Medical Biochemistry but is now run by the Swiss Institute of Bioinformatics (SIB). The information about protein sequences contains the explanation of the function of the proteins along with the information on the structure of the domains.

A Sample SWISS-PROT entry:

ID 112_LYCES STANDARD; PRT; 105 AA.

AC Q43495;

DT 15-JUL-1999 (Rel. 38, Created)

DT 15-JUL-1999 (Rel. 38, Last sequence update)

DT 15-JUL-1999 (Rel. 38, Last annotation update)

DE PROTEIN 108 PRECURSOR.

OS Lycopersicon esculentum (Tomato).

OC Eukaryota; Embryophyta;

```
OC   Spermatophyta; eudicotyledons;
OC   core eudicots; euasterids I; Solanales; Solanaceae;
OC   Solanum.
RN   [1]
RP   SEQUENCE FROM N.A.
RC   STRAIN=CV. VF36; TISSUE=ANTHER;
RX   MEDLINE; 94143497.
RA   CHEN R., SMITH A.G.;
RL   Plant Physiol. 101:1413-1413(1993).
DR   EMBL; Z14088; CAA78466.1; -.
DR   MENDEL; 8853; LYCes;1133;1.
KW   Signal.
FT   SIGNAL 1 40 POTENTIAL.
FT   CHAIN 41 106 PROTEIN 103.
FT   DISULFID 41 72 BY SIMILARITY.
FT   DISULFID 51 64 BY SIMILARITY.
FT   DISULFID 75 90 BY SIMILARITY.
FT   DISULFID 88 100 BY SIMILARITY.
SQ   SEQUENCE 105 AA; 1023 MW; AFA4231A CRC28;
     NASSTRVZPL KMMNGHTMNG LLIVFGTWSS ACVLLPKMNH JGVPDRTSSS LMNKJPPLMN
     PTGSSSTPLM NVASSTTMMN KLPPPTTRRW SSSTTPPMNV
//
```

The description is given below:

ID (IDentification line) line	is always the first line of an entry	
	Entryname	108_LYCES
	Dataclass	STANDARD;
	Molecule	PRT;
	sequence length (Amino Acids)	102 AA.
AC (Accession Number) line	gives the accession numbers which are connected with this entry.	
DT (DaTe) line	shows when an entry first appeared in the database and when it was last updated.	
DE (DEscription) lines	contain usual explanatory information about the sequence stored.	
OS (Organism Species) line	specifies the preferred scientific name of the organism which was the source of the stored sequence.	

The OC (Organism Classification) lines	contain the taxonomic classification of the source organism.
The RN (Reference Number) line	A sequential number to each reference mentioned in an entry.
RP (Reference Position) line	tells about the amount of the work carried out by the authors of the reference.
RC (Reference Comment) lines	store comments relevant to the reference cited. This is an optional field.
RX (Reference cross-reference) line	This is not a mandatory line and is used to tell about the identifier assigned to a specific reference in a bibliographic database.
The RA (Reference Author) lines	list the authors of the paper (or other work) cited.
The RL (Reference Location) lines	usually mention information for the reference.
DR (Database cross-Reference) lines	tell about the information of UniProtKB/Swiss-Prot entries and other information which is linked with collections of data other than UniProtKB/Swiss-Prot.
The KW (KeyWord)	This line provides information that can be used to create indexes of the sequence entries decided upon the functions, structures, or some further categories.
The FT (Feature Table) lines	FT (Feature Table) lines provides an easy way for sequence data comments. The table explains sites or regions of interest in the sequence.
The SQ (SeQuence header) line	has a line code consisting of two blanks rather than the two-letter codes which have been used uptil now. The sequence has 60 amino acids per line, in groups of 10 amino acids each, starting at position 6 of the line.
// (terminator) line	This has no data or comments and tells that this is the end of an entry.

3.9.5.2 PIR (Protein Information Resource)

The PIR had its origins in 1960 at the National Biomedical Research Foundation (NBRF). The Project was started to collect protein sequences to find the relationships among proteins that showed evolutionary links with each other. In 1988, the PIR-International took over the databases for maintenance. PIR-International is an organization which looks after the databases of proteins from different places. It particularly looks after 3 protein databases, namely, PIR (Protein Information Resource) at NRBF, JIPID (Japanese International Protein Sequence Database) and MIPS(Martinsried Institute for Protein Sequneces).

The PIR database is divided into 4 sections varying from PIR 1 to PIR 4. These sections are made so that the sequences can be split into four categories based on the quality of the data that they have and the amount of description each sequence contains.

1. **PIR 1 :** contains completely classified entries which have complete comments attached with them.
2. **PIR 2 :** contains first round of entries which have not been properly examined and may have unnecessary entries.
3. **PIR 3 :** contains entries which have not been examined at all and are therefore untested.
4. **PIR 4 :** conceptual translations of manmade sequences. These sequences are the ones which have not been transcribed or translated but have been made by the process of genetic engineering.

3.9.5.3 MIPS : Martinsried Institute for Protein Sequences

This database collects and processes data for the PIR-International database.

3.9.5.4 PDB (Protein Data Bank)

This database is mainly concerned in dealing with the molecular structure of molecules and also the raw data from where the molecular structures are obtained.

PDB offers three options for searching the database:

1. to enter the PDB identifier.
2. to use searchlite interface.
3. to use searchfield interface.

The searchlite interface allows you to enter the query by joining keywords like AND, OR and BUTNOT.

The serchfield is a customized interface in the sense you can filter your search by selecting different categories of search. Here you may search by author or FASTA search etc. therefore different search options are available to you to choose from. While using searchlite or searchfield, it provides you with an opportunity to reassess your query or print a report of your query.

3.9.5.5 Composite Protein Sequence Databases

The word composite means a set of primary databases for protein sequences. The composite database search yields better results since a number of databases are searched, but the same sequence is searched only once.

A number of composite databases are maintained such as NRDB, OWL, MIPSX and a combination of SWISS-PROT and TrEMBL.

3.9.5.6 NRDB (Non-Redundant Data Base)

The NCBI looks after this data base. The database is a composite of GenPept (a database which contains automatic translations of CDS (coding sequence: A region of DNA or RNA whose sequence determines the sequence of amino acids in a protein) from GenBank), PDB Sequences, SWISS-PROT, SPupdate (SWISS-PROT's weekly changes), PIR and GenPeptupdate (GenPept's daily changes).

The database is called non-redundant, but actually it is redundant since it contains multiple copies of the same proteins. But it tries to remove only the alike copies of the proteins. The entries in this database may also contain errors since wrong sequences that have been altered in SWISS-PROT are brought here again when they are retranslated from DNA sequences. This database is still the official database for the BLAST program.

3.9.5.7 OWL

This is also a non-redundant protein sequence database. It is a composite of:

1. SWISS-PROT
2. PIR(1-4)
3. GenBank (CDS Translations)
4. NRL-3D

These four sources are given a precedence number based on the level of confirmations present in their sequences. SWISS-PROT gets the highest precedence. Any new entry which is to be added to this database is first matched with SWISS-PROT, if it is present in SWISS-PROT it is not added otherwise added to the database.

3.9.6 Secondary Databases

Secondary databases use the primary databases as their source to build information which is based on the inferences drawn out of the analyses carried out on the primary sources. All of the secondary databases contain different types of information.

The way to build secondary databases out of primary sources is to first look for data in the primary sources then to study the data to understand the results and finally to find if there is any biologically important meaning attached to it.

The most common primary resource used by many secondary databases is the SWISS-PROT.

Some of the main secondary databases are given below along with the description of the information they hold and the name of the primary database they use for taking the data from.

S.No	Name of secondary database	Name of primary resource	Type of information held	Type of information commonly known as
1	PROSITE	SWISS-PROT	Regular Expressions	Patterns
2	Profiles	SWISS-PROT	Weighted matrices	Profiles
3	PRINTS	OWL	Aligned motifs	Fingerprints
4	Pfam	SWISS-PROT	Hidden Markov Models	HMMs
5	BLOCKS	PROSITE/PRINTS	Aligned motifs	Blocks
6	IDENTIFY	BLOCKS/PRINTS	Fuzzy Regular Expressions	Patterns

As it can be seen from the above table, the information stored in all of these databases is different but they all have a universal basis to develop from namely the concept of homologues. Homologous sequences are those sequences which differ in some respect but have a common ancestor. These homologous sequences are gathered in multiple alignments and some of the regions in these sequences are conserved in the sense that they show little or no difference. These conserved regions are called motifs which generally have some biologically important meaning attached to them, i.e. they show similarity in either the function or structure of the protein.

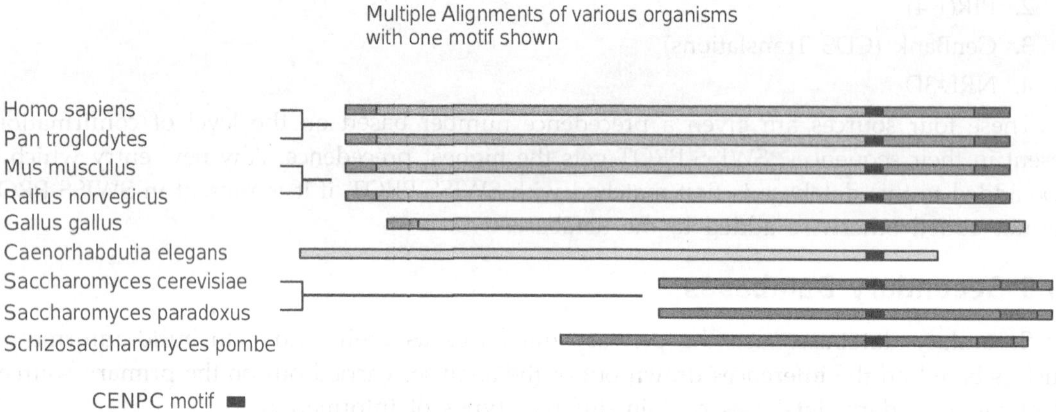

Fig. 3.2: Multiple Alignments

These motifs help to identify protein families. The motifs play an important role in building the secondary databases. Let us describe how each one of these secondary databases is built.

3.9.6.1 Prosite

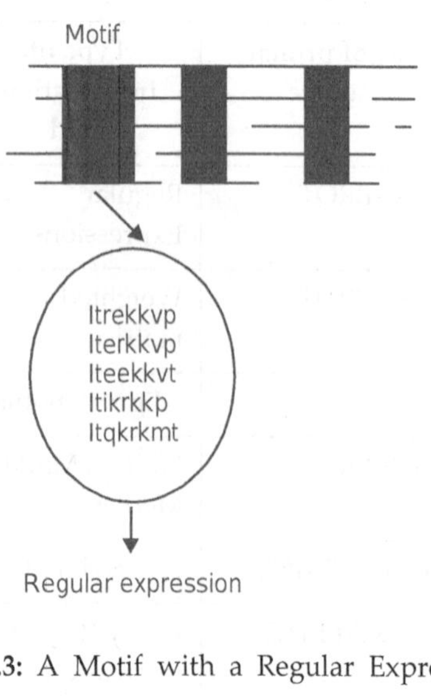

Fig. 3.3: A Motif with a Regular Expression

PROSITE was the first secondary database that came into existence and was maintained by the Swiss Institute of Bioinformatics. The multiple alignments of known homologues (common ancestors) form a motif which is the most conserved region among the alignments. The newly generated sequence is tested against this motif to find whether the sequence belongs to the protein family which this motif represents. In such a way, the sequence is compared with a number of motifs to find out to which protein family the sequence belongs.

The motifs are stored within PROSITE in the form of regular expressions, which are also called patterns. The way these regular expressions are built is by first generating the multiple alignments and then manually looking for conserved regions. The motifs contain information from multiple sequences and later all these sequences are reduced to a single expression.

The pattern formed was taken by comparing the sequences among a number of entries from the SWISS-PROT and later when the pattern is formed, it is again compared to the SWISS-PROT entries to see how the pattern is performing.

A Sample PROSITE entry

```
ID   PEPTIDASE_U12; PATTERN.
AC   PS01276;
DT   OCT-1995 (CREATED); OCT-1995 (DATA UPDATE); MAY-2009 (INFO UPDATE).
DE   Peptidase family U32 signature.
PA   E-x-F-x(2)-G-[SA]-[LIVM]-C-x(4)-G-x-C-x-[LIVM]-S.
NR   /RELEASE=57.3,468851;
NR   /TOTAL=10(10); /POSITIVE=10(10); /UNKNOWN=0(0); /FALSE_POS=0(0);
NR   /FALSE_NEG=1; /PARTIAL=0;
CC   /TAXO-RANGE=A??P?; /MAX-REPEAT=1;
CC   /VERSION=1;
DR   P59916, PRTC1_PORGI, T; P33437, PRTC2_PORGI, T; Q57555, Y090_METJA , T;
DR   Q9ZMR3, Y155_HELPJ , T; P56113, Y169_HELPY , T; P44700, Y419_HAEIN , T;
DR   P76104, YDCP_ECOLI , T; P76403, YEGQ_ECOLI , T; P45527, YHBU_ECOLI , T;
DR   O32034, YRRO_BACSU , T;
DR   Q59051, Y1657_METJA, N;
DO   PDOC00982;
//
```

The description is given below:

ID	is the name given to the Family . Pattern word indicates that it is a regular expression.
AC	This is the accession number connected with this entry.
DT	shows when an entry first appeared in the database and when it was last updated.
DE	This tells about the general explanatory information about the protein family.
PA	specifies the pattern.
NR	Gives the details as to how pattern was formed.
CC	Comments
DR	Contains the number of matches found. T for TRUE matches and P for possible matches. False positives are indicated by F and False Negatives by N.
DO	Gives the documentation file related to the protein family.
// (terminator)	This field does not contain any data or comments and tells that the entry has ended.

3.9.6.2 Prints

The PRINTS database mainly holds signatures to the family of proteins. The signatures are conserved motifs themselves but it is not a single conserved motif but a number of conserved motifs are taken together to form a fingerprint which becomes the signature to identify the family of proteins. The fingerprints provide more trustworthiness than the patterns since in patterns only one conserved motif is used but here a number of them are used.

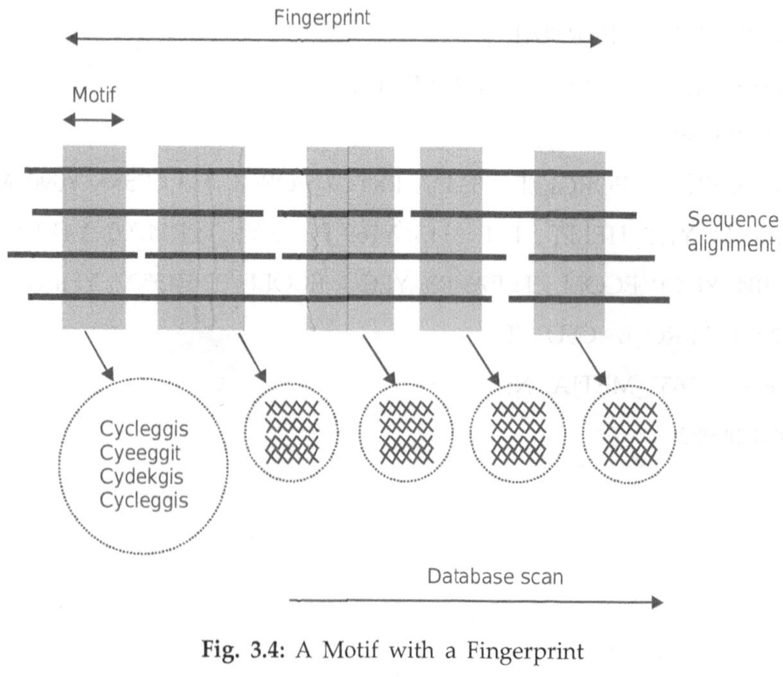

Fig. 3.4: A Motif with a Fingerprint

In the PRINTS database, the fingerprint motifs are formed in a different way. Firstly the initial multiple alignments of all the motifs of a fingerprint are taken, then the database is searched for sequences that match the motifs alone. If there is information in these other sequences which is not present in the initial multiple alignment sequences, then this new information from these new sequences is added to these motifs and the database is searched again. The process is done again until no more sequences which match the motif are found and this is how conserved motifs are increased and the results added to the database.

A Sample PRINTS entry

```
gc; BACTRLOPSIN
gx; PR00431
gn; COMPOUND(7)
ga; 1-OCT-1990; UPDATE 16-DEC-1996
gt; Bacterial opsin signature
gp; PRINTS; PR00399 GPCRRHODOPSN; PR00402 GPCRCAMP; PR00421 GPCRMGR
gp; PRINTS; PR00445 GPCRSECRETIN; PR00478 GPCRSTE4; PR00599 GPCRSTE6
gp; PRINTS; PR00400 OPSIN
gp; INTERPRO; IPR001211
gp; PROSITE; PS00987 BACTERIAL_OPSIN_1; PS00354 BACTERIAL_OPSIN_RET
gp; BLOCKS; BL00344
gp; PFAM; PF01432 Bac_rhodopsin
gp; PDB; 1BRD
gp; SCOP; 1BRD
bb;
gr; 1. GATE, M. AND CHANG, L.
gr; one pump, one principle - membrane-transport in bacteria.
gr; TRENDS BIOINFOMATICS.SCI. 11(1) 88-99 (1986).
gr;
gr; 2. GEETHA, M., MURLI, T., CHANG, J., POTTER, R.AND SWAMY, A.
gr; Primary structure of rhodopsin-2, a prokaryotic photoreceptor.
gr; EMBRYO J. (14) 1123-1145 (1986).
gr;
gr; 3. NATH, L., JHA, P. AND SATYA, T.
gr; Properties and the primary structure of a new photoreceptor
from
```

```
gr; strain mex.
gr; BIOCHEMISTRY.BIOPHYSICS 1134 45-56(1989).
bb;
bb;
gd; The bacterial opsins are proteins that provide light-
gd; dependent membrane transport functions to a family of
gd; bacteria [1,2].
gd; They are integral membrane proteins believed to contain
gd; 9 transmembrane (TM) domains

si; SUMMARY INFORMATION
si; --------------------
sd; 12 codes involving  9 elements
sd; 2 codes involving   7 elements
sd; 2 codes involving   5 elements
sd; 0 codes involving   2 elements
bb;
bb;
ci; COMPOSITE FINGERPRINT INDEX
ci; ---------------------------
cr;
cd;  9|  12  12  12  12  12  12  12
cd;  7|   0   0   2   2   2   2   2
cd;  5|   0   2   2   0   0   2   2
cd;  4|   0   1   1   0   0   1   1
cd;  3|   0   0   1   0   1   1   1
cd;  2|   0   0   0   0   0   0   0
cd; --+------------------------------------
cd;   |   1   2   3   4   5   6   7
bb;
bb;
bb;
```

```
im; INITIAL MOTIF-SETS
im; ------------------
ic; BACTRLOPSIN1
il; 18
it; Bacterial opsin TM motif I - 1
id; IMYLIMNGRSDCTRANMTRA   BAC2_HALOP   22   22
id; AYMNLKHYTRACNMLTRDGH   BACH_NATLA    7    7
id; GLMKNGYRADSCLPQRTGHA   BACS_HALSO   12   12
id; LWSHYTGCBDTGLPWAQQRT   BAC1_HALSP   18   18
id; IWGRTASMCDSTRQPOLKHG   BACR_NATPH   19   19
bb;
ic; BACTRLOPSIN2
il; 19
it; Bacterial opsin TM motif II - 1
id; WRDSCRSSTGBCKLPTOPLL   BAC2_HALOP   55   10
id; AYWRSHLTOPMNGRSTWGHT   BACH_NATLA   34   12
id; HJLPORWQSANMCFTGYPRR   BACS_HALSO   51   10
id; HKLOPRWQSABMKLOPNCDF   BAC1_HALSP   48   13
id; STRAFGHPOLTRGCFSHGTR   BACR_NATPH   49   13
bb;
ic; BACTRLOPSIN3
il; 20
it; Bacterial opsin TM motif III - 1
id; WRGHDSTYPOLKMNSDRGHT   BAC2_HALOP  110   39
id; WDSRTGHLKOPMNDRTSWAA   BACH_NATLA   60   10
id; GHTRDSWAQOPLMNGRDSGH   BACS_HALSO   98   21
id; GHTROPMNGOLPYGMNLPFG   BAC1_HALSP   85   15
id; FRTWQPOLKPOLTRWSNMNM   BACR_NATPH   86   15
bb;
```

Description of the above:

Gc	The title of the family
Gx	The accession number
Gn	The number of Motifs
Ga	The date it was created and modified
Gt	The word signature indicates that it is a fingerprint.
Gp	Links to other related databases
Gr	Bibliographic information
Gd	Technical details regarding the fingerprint
Si	Summary information
Sd	Summary details. It indicates that 15 sequences have matched all 7 motifs, 1 sequence has matched 6 out of 7 motifs etc.
Ci	Composite index
Cd	Composite details. The same information is shown in a composite manner.
Im	Initial Motifs are given
Ic	Name given to each motif
Il	Length of each motif
It	Iteration number of the motif
Id	Description of the sequence along with organism, the first number shows the starting position in the parent fragment. The second number shows the number of residues between the fragment and its preceding neighbour. In the first motif it equals to the starting location of the fragment.
Fm	Final Motifs
Ft	Final iteration
Fd	Final description

3.9.6.3 Blocks

Blocks, or motifs, match to minimal units of protein function. They are short amino acid sections that are conserved in sequence and in length. Although individual amino acids contain smaller units than blocks, they are not enough to define a unique function.

Block-based alignment methods differ from those based on global multiple sequence alignment. Both perform better than single sequence analyses in identifying the functionally critical sequence regions from a group of related sequences. Block-based methods are explicitly designed to identify conserved regions, whereas more global multiple sequence alignment usually includes alignment of both conserved and non-conserved regions.

Making blocks: Blocks are formed by the automated PROTOMAT system which applies a robust motif-finder to a set of interrelated sequences of protein. The consequential candidate motifs are put together into a best set along the lengths of the sequences to give a multiple alignment consisting of ungapped conserved regions separated by unaligned regions of variable size.

MOTIF looks for spaced triples in most of the sequences and aligns them around these triples. A spaced triple is a set of three amino acids separated by two distances. For Block Maker, all spaced triples with all combinations of two distances ranging from between 0 and 17 amino acids each are tallied.

Searching with blocks: Block alignments are converted into position-specific scoring matrices (PSSMs) for searching. Each PSSM column corresponds to a block position and includes 20 numerical scores representing the odds for each amino acid occurring in that position. A theoretical score distribution is computed for each PSSM. A query sequence is compared with each PSSM in the Blocks Database by aligning it with the block at every possible position and adding the log-odds scores in each PSSM column. The highest-scoring alignment is saved and the probability of its score looked up in the theoretical distribution. For families with multiple blocks, each block is aligned and scored individually with the query, and the probabilities of all the blocks are combined to give the overall expected value for the alignment of the query with the blocks for the family. Multiple blocks are only combined in a hit if they occur in order and within reasonable distances of one another within the query sequence. Reasonable distances are determined by looking at the distances between blocks in the known members of a family.

3.9.6.4 Profiles

A different method than the motif based method of protein family takes into account the fact that different regions between conserved motifs also contain different type of information. Hence to differentiate among two sequences, there are the whole sequences which become the differentiators. This differentiator which is known as a profile is measured to indicate where the insertions and deletions can occur. At what place what kind of residue can take place and where are the very much conserved regions.

Profiles are also known as weighted matrices provide a discrete way of finding out far-off associations, where only a few residues are the same and the rest are different yet they do hold some meaning and that is important.

A Sample PROFILES entry

```
ID  C1;  MATRIX.
AC  PS20005;
DT  OCT-1992  (CREATED);  OCT-1992  (DATA  UPDATE);  JUN-2002  (INFO
UPDATE).
DE  C1  domain  profile.
MA  /GENERAL_SPEC:  ALPHABET='ABCDEFGHIKLMNPQRSTVWYZ';  LENGTH=129;
TOPOLOGY=LINEAR;
MA  /DISJOINT:  DEFINITION=PROTECT;  N1=2;  N2=121;
MA  /NORMALIZATION:  MODE=1;  FUNCTION=LINEAR;  R1=0.1332;  R2=0.0231;
TEXT='-LogE';
MA  /CUT_OFF:  LEVEL=0;  SCORE=344;  N_SCORE=6.6;  MODE=1;  TEXT='!';
MA  /CUT_OFF:  LEVEL=-1;  SCORE=355;  N_SCORE=5.8;  MODE=1;  TEXT='?';
MA  /DEFAULT:  B1=-80;  E1=-40;  MI=-128;  MD=-128;  IM=-128;  DM=-128;
I=-70;  D=-10;
```

```
MA /I: B1=0; BI=-128;
MA /M: SY='L'; M=-12,14,-34,50,17,-32,-8,1,-22,1,-22,-24,21,-8,1,-6,
2,-5,-14,-34,-21,14;
MA /M: SY='K'; M=-5,-15,-18,-9,-2,-24,-14,-23,-14,-7,-15,-16,-8,67,-9,
-43,-6,-5,-12,-15,-43,-9;
. .
MA /M: SY='T'; M=-3,-5,-6,-8,-10,-23,-32,-23,-3,-5,-7,-9,-4,-13,-1,
-2,1,8,-2,-14,-5,-2;
MA /I: E1=0; IE=-128; DE=-128;
NR /RELEASE=43.2,3445564;
NR /TOTAL=342(202); /POSITIVE=342(202); /UNKNOWN=0(0); /FALSE_POS=0(0);
NR /FALSE_NEG=177; /PARTIAL=0;
CC /MATRIX_TYPE=protein_domain;
CC /SCALING_DB=reversed;
CC /AUTHOR=G_SANTHI;
CC /TAXO-RANGE=??E??; /MAX-REPEAT=4;
CC /FT_KEY=DOMAIN; /FT_DESC=C2;
CC /VERSION=1;
DR A5QNS1, ABR_BOVIN , T; Q13429, ABR_HUMAN , T; Q345G4, ABR_MOUSE
, T;
DR Q43252, ABR_XENLA , T; A54r44 ABR_XENTR , T; Q3S421, ADCB_DICDI
, T;
DR Q542525, ADCC_DICDI , T;QR5542 AGD11_ARATH, T;Q34R54 AGD12_ARATH,
T;
DR Q4324G5, AGD13_ARATH, T;O98766, BAIP3_HUMAN, T;R654T5, BAIP3_MOUSE,
T;
DR P26261, BCR_HUMAN , T;Q665J1, BCR_MOUSE , T; P13224, BUD2_YEAST
, T;

. . .

DR Q25522, WWP2_MOUSE , N;Q33535, Y0670_DICDI, N;O98827, YDY2_SCHPO
, N;
PR PRU00032
DO PDOC00232;
//
```

The description is given below:

ID	The word matrix tells that it is a profile.
MA	Line provides various details of the residues given by the alphabet and the cut off scores that they yield.
CC	is comments
DR	is database references showing true positives and true negatives.

3.9.6.5 Pfam

Pfam is a database of protein families that includes their comments and multiple sequence alignments generated using hidden Markov models.(A hidden Markov model (HMM) is a statistical model in which the system being modelled is assumed to be a Markov process with unobserved state). For each family in Pfam you can:

- look at multiple alignments
- view protein domain architectures
- view known protein structures
- The Pfam database contains information about protein families and protein domains (A protein domain is a part of protein sequence and structure that can develop gradually, work, and occur independently of the rest of the protein chain. Each domain makes a packed three-dimensional structure which most of the times can be folded independently and can be stable on its own. Many proteins consist of several structural domains. One domain can show up in a number of proteins that are evolutionarily associated. Domains differ in length say from 25 amino acids up to 500 amino acids). Pfam-A contains over 9,000 entries, which are checked using a manual alignment editor. For each entry a protein sequence alignment and a hidden Markov model is stored. Pfam-A entries do not deal with all known proteins and therefore Pfam_B is created which is an automatically generated supplement. Although of lower quality, Pfam-B families can be useful when no Pfam-A families are found.

An Excerpt from a Pfam entry

```
ID 6tm_2
AC PF00002.12
DE 5 transmembrane sweet-taste receptor of 2 GCPR
AU Sonnhammer ELL
SE Prosite
GA -102.00 -102.00; 12.60 12.60;
TC -101.40 -101.40; 12.80 12.80;
NC -105.90 -105.90; 11.30 11.30;
TP Family
BM hmmbuild -FHMM_ls.ann SEED.ann
BM hmmcalibrate --seed 0 HMM_ls
```

```
BM hmmbuild -f -FHMM_fs.ann SEED.ann
BM hmmcalibrate --seed 0 HMM_fs
AM byscore
RN [1]
RM 16527728
RT Identification of the interaction site within the
RT domain of the human sweet-taste receptor
RT T1R2.
RA MURLI R, CHANG J, SHANTI B, POTTER A, SHAW Q,
RA JAMES RF;
RL J Biological Chemistry. 2002;210:12345-12566.
DR INTERPRO; IPR000221;
DR PROSITE; PDOC00231;
CC This is a domain of five transmembrane regions that form the
CC 2 protein receptors which are grouped as G-coupled. These are
CC often associated with a linker domain,
CC NCD3G Pfam PF035443, which is the human sweet-taste receptor,
CC and ANF_receptor Pfam:PF190991.
CC [1].
SQ 5232
Q2UNX6_ICTPU/213-345     AC Q2UNX6.1
O23445_FUGRU/2-112       AC O23445.1
Q34511_AEDAE/281-299     AC Q34511.1
Q2WKZ2_DROOR/322-422     AC Q2WKZ2.1
. . .
//
```

Description of the above:

ID	A single keyword
AC	Accession number
AU	Author
DE	Description
SE	Seed Database
DR	Database Cross Reference
GA	Search Program Cut offs used to build full alignments.
CC	Comments
SQ	Sequences

3.9.6.6 Identify

The content of IDENTIFY databases are fuzzy regular expressions. These differ from the regular expressions of the PROSITE database in the sense that little fuzziness is incorporated in the regular expressions. In other words, these fuzzy regular expressions not only match exact amino acids but match with other amino acids also which show similar biochemical properties such as charge, size and others.

Since this method does not strictly match the amino acids, it introduces scope for noise. But on the other hand, if little flexibility is allowed, then significance attached to minority matches can be found out.

Here is a table which shows how different groups can be adjusted together based on the similar properties that they possess.

Property	Groups
Small	Ala, Gly
Small hydroxyl	Ser, Thr
Basic	Lys, Arg
Aromatic	Phe, Tyr, Trp
Small Hydrophobic	Val, Leu, Ile
Medium Hydrophobic	Val, Leu, Ilc, Met
Acidic/amide	Asp, Glu, Asn, Gln
Small/polar	Ala, Gly, Ser, Thr, Pro

The names given in the right hand column of the above table are some of the names of amino acids out of the 20 amino acids.

3.9.7 Structure Classification Databases

This database holds the 3D structures of proteins. In our earlier discussion, we talked about X-ray crystallography, so from those experiments and other studies, the structures can be easily obtained. Most proteins have structures which are similar to each other and those proteins are likely to have common ancestors. During the process of evolution the amino acid sequences get deviated down the chain, due to insertions, deletions and substitutions. Due to these changes the folds of the proteins become different and large and as a result their structure changes but their function remains the same. Thus, there exist a relationship between the structure and sequence of amino acids, which is better understood by a variety of structural classification schemes.

Some classification schemes use automatic inspection while other rely on automatic procedures.

There are two well known classification schemes known as SCOP classification and CATH classification. Let us briefly describe them.

3.9.7.1 SCOP(Structural Classification Of Proteins)

The SCOP database has data related to proteins with known structure and holds information regarding the structural and interesting results that arise due to the evolution (evolution is the change in the genetic material of a population of organisms from one generation to the next. Although the differences generated in whichever generation are small, changes are built up with each new generation and may, with due course of time, be able to make significant differences in the organisms) process.

SCOP makes use of 4 different levels of classification which are in a hierarchical structure:

1. **Class:** General "structural architecture" of the domain.
2. **Fold:** Similar arrangement of regular secondary structures but may or may not have common evolutionary origin.
3. **Superfamily:** Sufficient structural and functional similarity to infer a common evolutionary origin but not necessarily show identity in sequences.
4. **Family:** Some sequence similarity can be detected.

This classification is more significantly based on the human expertise. It is usually accepted that SCOP provides a better justified classification. Human expertise is needed to decide whether certain proteins are evolutionary related and therefore should be assigned to the superfamily which is equal, or their resemblance is an outcome of structural constraints and therefore they belong to the same fold.

3.9.7.2 CATH (Class, Architecture, Topology, Homology)

The CATH Protein Structure Classification is a partly self operated, consisting multiple levels of classification of protein areas published in 1997. The name CATH is an acronym of the four main levels in the classification.

The 4 major levels of the CATH hierarchy are as follows:

S.No.	Level	Description
1	Class	By enlarge contents of the secondary-structure and packing of the domain.
2	Architecture	A significant combination of topologies which have specific structural features. Secondary structure arrangement matches but not connectivity.
3	Topology	High structural similarity but no evidence of homology. Overall shape and connectivity of secondary structure matches. Equivalent to a fold in SCOP
4	Homologous superfamily	Indicative of a demonstrable evolutionary relationship, i.e. show identity and have a common ancestor. Equivalent to the superfamily level of SCOP.

CATH also defines 4 classes: which are called mostly-alpha, mostly-beta, alpha and beta, few secondary structures.

In the CATH classification system much of the work is done by automatic methods, however there are important manual processes to do the classification.

The first step is to separate the proteins into domains.

The domains are sorted on their own into classes and based on the similarity of sequences they are clustered together. These groups form the H levels of the classification. The level of topology is made by comparing structures of the homologous groups. Lastly, the level of the Architecture is given manually.

Class Level classification is done on the basis of 4 criteria.

(a) Secondary structure content

(b) Secondary structure contacts

(c) Secondary structure alternation score

(d) Percentage of parallel strands.

3.9.7.3 PDBsum

PDBsum is a web-based database providing a large summary of pictures of the information on each structure given in the Protein Data Bank (PDB). It contains images of structures, explaining secondary structure of each protein chain, complete structural analysis created by the PROMOTIF program, abstract of PROCHECK results and graphic diagrams of protein-ligand and protein-DNA interactions. Scripts of RasMol emphasize main features of the structure, which may be the domain of the proteins, or patterns of the PROSITE database or can be protein-ligand interactions, which can be viewed in 3D format with the facility to interact with it.

3.10 THE QUALITY OF INFORMATION

The quality of data or information can be evaluated by the three parameters given below:

1. Author of the Information
2. Authenticity of Information
3. Time of Publishing the Information

By Author of information we mean that who is the one who is publishing the website. What is the intension of the corporation funding the information? Is the corporation a government body, a research body or a profit making company? Why does it make the information available to the casual user? Does the corporation answers to the user's queries? What is the detail level to which the information is reached? Does the information holds errors? GenBank is assumed to have one error per sequence. The NCBI and PDB are supposed to have good quality data.

By Authenticity of information we mean does the website in consideration provide enough documentation and links to the sources from where the data is built. Does it allow the user to read the source code and provide cross reference links to bibliographic information and referred journals. If such information is provided then we can assume that the information is authentic and not just anyone has made the website.

By Time of Publishing the Information we mean that there are many websites which were developed earlier and worked well with the technologies and libraries printed then. But these websites still exist on the server are still use the earlier technologies and libraries. There may be problems in using these technologies and libraries which do not carry the latest information.

3.11 METADATA

Metadata is a term used in connection with data and information. Let us briefly define the three terms together and then show with the help of an example, the difference between them.

1. Data : Data are raw figures or facts derived from check up, examination, measurement and computation.
2. Information : Processed Data, concerned statements and analysis of a specific object, event or process.
3. Metadata : Data about the reference in which information has been made use of. Explanatory summaries or grouping of data and information at a greater plane.

Example:

Data : Patient age–15 yrs, fever, cough, sore throat, headache, running nose.

Information : Swine flu is a disorder of the respiratory system in pigs caused by type A influenza virus. In humans some people show signs of vomiting and diarrhea.

Metadata : In 1918, the flu effected one third of the world's population and caused 50 million deaths. In 2009 the flu has so far reported to have effected 1.2 million people and caused 700 deaths. Generally the flu can be treated and has a low death rate.

Unit 4

MOLECULAR BIOLOGY AND BIOINFORMATICS

4.1 CENTRAL DOGMA OF MOLECULAR BIOLOGY

The central dogma of life or the central dogma of molecular biology consists mainly of three processes, which are replication then transcription and then translation.

The central dogma of molecular biology was first described by James Watson. He said, the process of replication duplicates DNA, the process of transcription converts a DNA into RNA and the process of translation converts RNA into proteins.

The Genome is an individual's genetic information which is encoded as a sequence of nucleotides (adenine, guanine, cytosine and thiamine) on a DNA strand.

In human cells, DNA is present in the form of 23 pairs of chromosomes, with each one of the pair coming from each one of the parent. It is commonly known that around 98.5% of the function of DNA is of no use and is considered as garbage. The rest of the DNA is what we call genes and these are the ones that are concerned with the process of transcription.

In the DNA sequence, the base pairs that are expressed are known as *exons* and the ones that are not expressed are known as the *introns*.

During the process of transcription, precursor mRNA (pre-mRNA) is created from the DNA, which is an undeveloped single strand of messenger RNA (m-RNA). It is also known as heterogeneous nuclear RNA (hnRNA). When the hnRNA gets processed, it is converted into mature messenger RNA or mature mRNA or simply commonly known as mRNA. Pre-mRNAs include two different types of segments, exons and introns. The introns are removed from the final mRNA by the process of splicing and the exons are the ones that exist in the final mRNA.

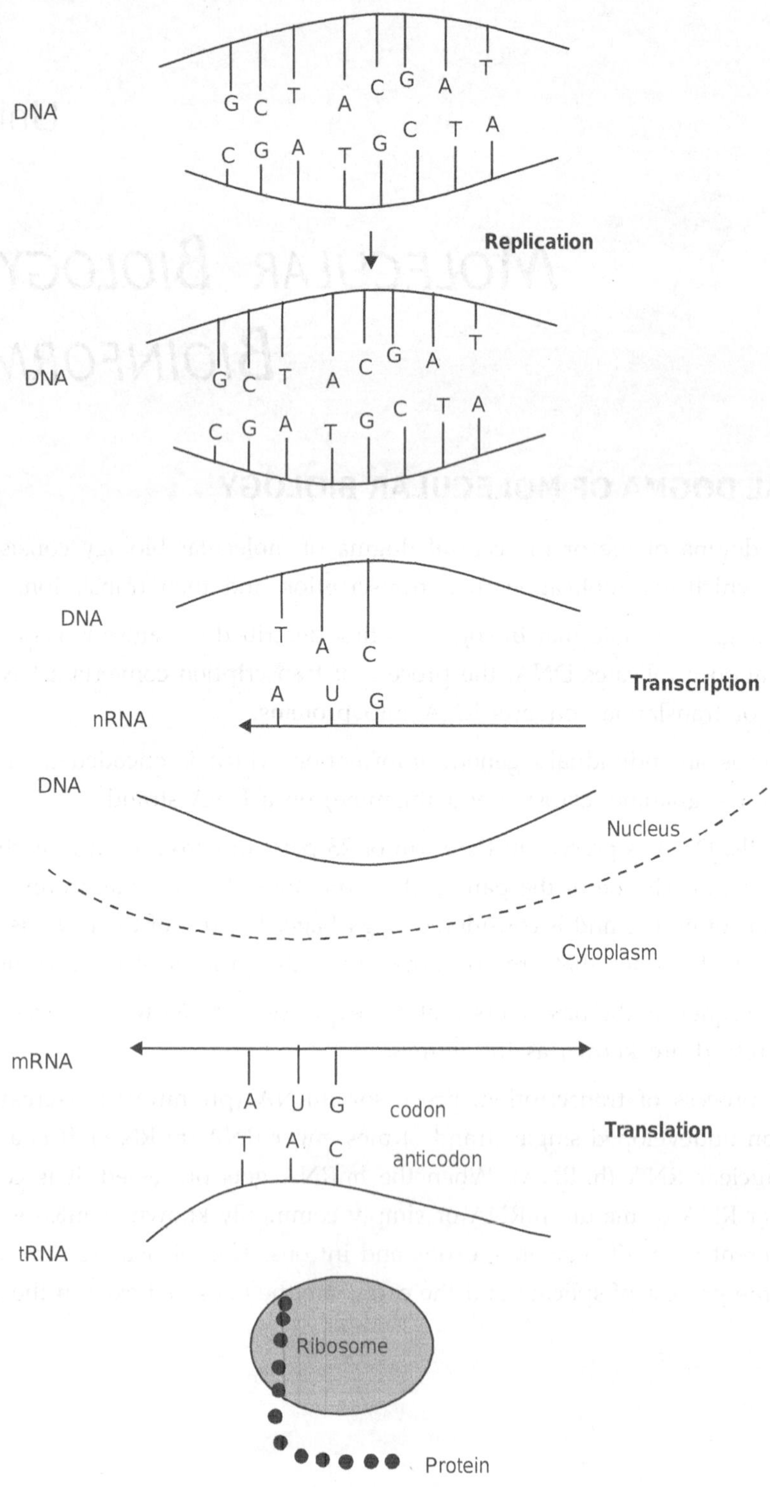

Fig. 4.1: The Central Dogma of Molecular Biology

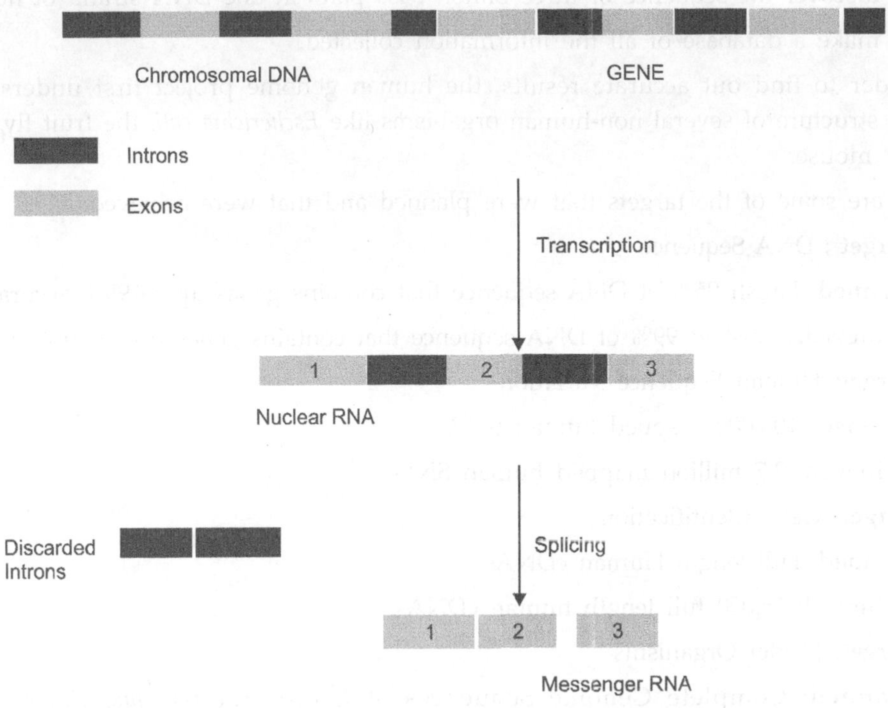

Fig. 4.2: Conversion of DNA into mRNA

The mature mRNA is transported through the nuclear membrane into the cytoplasm where the process of translation converts the mRNA into proteins with the help of ribosomes.

These ribosomes contain a collection of RNA molecules known as ribosomal RNA or rRNA. The transient rRNAs help in attaching the mRNA to the ribosome when the translation is taking place. Another type of RNA which helps in translation is the transfer RNA or tRNA which carries an anticodon and attaches that to the codon. The codon is a three letter base sequence (ACGT) on the mRNA strand. The anticodon is the three letter alphabet which is associated with 20 amino acids of proteins.

During translation, amino acids are appended to the linear chain of proteins and the ribosomes move from one codon to another codon on the mRNA strand.

4.2 THE HUMAN GENOME PROJECT

The human genome project was started by the U.S. Department of Energy in collaboration with the National Institute of Health in the year 1990 and was completed in 2003.

Around 18 countries have taken HGP initiative. They are, Australia, Brazil, Canada, China, Denmark, European Union, France, Germany, Israel, Italy, Japan, Korea, Mexico, Netherlands, Russia, Sweden, United Kingdom, and the United States.

The goals of the project included:

1. to categorize all the genes (20,000-25,000) in the DNA of humans.

2. to discover the sequence of three billion base pairs in one DNA strand of humans.
3. to make a database of all the information collected.

In order to find out accurate results, the human genome project first understood the hereditary structure of several non-human organisms like *Escherichia coli,* the fruit fly, and the laboratory mouse.

Here are some of the targets that were planned and that were achieved:

1. **Target :** DNA Sequence

   Planned: Finish 95% of DNA sequence that contains genes up to 99% accuracy.

   Achieved: Finished 99% of DNA sequence that contains genes up to 99% accuracy.

2. **Target:** Human Sequence Variation

   Planned: 100,000 mapped human SNPs

   Achieved: 3.7 million mapped human SNPs

3. **Target:** Gene Identification

   Planned: Full length Human cDNAs

   Achieved: 15,000 full length human cDNAs

4. **Target:** Model Organisms

   Planned: Complete Genome Sequences of *E. coli, S. cerevisiae, C. elegans,* and *D. Melanogaster*.

   Achieved: Finished genome sequences of *E. coli, S. cerevisiae, C. elegans, D. melanogaster,* and also of others like *C. briggsae, D. pseudoobscura,* mouse and rat.

The scientists accumulated blood (female) or sperm (male) samples from a huge variety of donors. They worked only on a small fraction of samples and made them the DNA sources. The donors' identities were guarded and even the scientists didn't come to know about it. Otherwise many DNA clones were taken from already stored libraries.

The human genome project was to be made available to public with the help of the following databases:

1. NCBI Human Genome Sequencing Progress
2. University of California located in Santa Cruz
3. Ensemble Genome Browser at Sanger Institute and EMBL's European Bioinformatics Institute.
4. EMBL-Bank – EMBL's Nucleotide Sequence Database

The human genome project was useful in many areas like:

1. Medicine
2. Forensic Science
3. Agriculture
4. Evolution of humans
5. Health damage

In medicine it helped to:

(a) find out genetic links as to why diseases like neurofibromatosis, Alzheimer's disease, and familial breast cancer occur.

(b) find out in general the causes of diseases and diagnosis can be done at an earlier stage and diseases treated at an earlier stage.

(c) drugs can be designed according to the genetic make up of a person.

(d) damaged genes may be removed and new genes may be added via gene therapy.

(e) link organs that can be given for donation with the genetic make up of the persons receiving them in transplantation of organs.

In forensic science it helped to:

(a) detect the individual humans who may be associated with the crime by inspecting the DNA sequences.

(b) confirm paternity relationships.

(c) let those people go who have been wrongly arrested for the crime.

In agriculture it helped to:

(a) give people more healthy food, which is also pesticide free.

(b) crops grown can be insect and drought tolerant.

(c) increase the growth of crops.

(d) decrease the loss in crops.

In evolution it helped to:

(a) find out the evolution of humans.

(b) compare humans with other organisms and find out the genes that are responsible for different diseases and disorders.

(c) study the Y chromosome of males and its effect in evolution.

(d) study the X chromosome in females and its effect in evolution.

In health Damage it helped to:

(a) find out how some people get more affected by the same amount of radiations and others get less affected.

(b) find out the cancer causing toxins and their relations to different genetic make ups of individuals.

(c) understand mutations.

How was the genome sequencing done is described below.

1. Chromosomes, which are about 50 million to 250 million base pairs, had been cut into smaller pieces.
2. The smaller piece was then taken as a template and fragments were created that were different from each other in size by only one base pair.

3. The fragments were then separated by gel electrophoresis.
4. The final base at the end of each fragment was recognized. This procedure recreated the original sequence of As, Ts, Cs, and Gs for each small piece generated in first step.

The observations from the project were:

1. The human genome is made up of 3164.7 million chemical nucleotide bases (A, C, T, and G).
2. The largest gene has 2.4 million base pairs but on an average a gene consists of about 3000 base pairs.
3. All humans have about 99.99% similar genomes.
4. We still do not know the functions of about 50% of the genes that we have discovered.
5. The proteins are made only from 2% of the entire genome.
6. Repeated sequences which do not help in the making of proteins, also called "junk DNA" make up about 50% of the entire genome.
7. This "junk DNA" seems to be responsible for providing structural support to the chromosomes.
8. The regions of DNA which have genes are mainly composed of C and G nucleotides and the regions which do not have genes are mostly made up of A and T nucleotides.
9. Chromosome 1 has the highest number of genes (2968), and the Y chromosome has the least number of genes (231).
10. Other organisms have their genes uniformly spaced in comparison to the genes of a human which are randomly spaced.
11. The human genome has 50% of repeated sequences in comparison to the worms or flies which have 7% and 3% respectively.
12. There are about 1.4 million regions in the human genome which show differences in single base DNA (SNPs).

The cloning process:

Cloning means producing the genetic twin of an organism. First we should understand that three types of cloning exists and the differences between them: (1) DNA cloning, (2) reproductive cloning, and (3) therapeutic cloning.

DNA cloning is also known as molecular cloning or gene cloning. The gene which is of interest is cloned to make multiple copies of it for further analysis. Plasmids are self-replicating extra-chromosomal circular DNA molecules.

The gene is cloned by taking a DNA fragment which has the gene and removed from the chromosomal DNA with the help of an enzyme which cuts the DNA also known as restriction enzyme and then placed with a plasmid also known as a cloning vector which was cut with the same restriction enzyme. The plasmid produces multiple copies.

Reproductive cloning is to produce an animal that has the same nuclear DNA as an existing animal or a previously existing animal. Dolly was a sheep which was successfully produced after this process. This process is also known as "somatic cell nuclear transfer" (SCNT). In this method an adult cell's hereditary material which was the donor was removed from the nucleus and put in a recipient's egg. The recipient's egg had already got its hereditary material removed to make space for the new one. The egg then treated with chemicals to initiate the process of cell division. Once the cloned embryo is of appropriate size, it is placed inside the uterus of a female host and stays there until birth is given.

Therapeutic cloning is to produce a human clone for research. The intent is not to produce similar human beings but to develop stem cells to cure diseases. Stem cells are certain types of cells which can repair themselves by cell division and also can generate any type of specialized cells in the human body. Stem cells are removed from an egg which has divided for 5 days. The embryo dies after the process of extraction.

DNA cloning is mainly useful in genetic engineering, gene therapy and genome sequencing experiments. Genetic engineering mainly deals with making healthier crops, gene therapy helps to cure diseases and genome sequencing is to generate new genomes from DNA fragments.

Reproductive cloning is helpful in producing animals which are of use to humans, for example producing animals that produce some drugs, or animals on which experiments can be conducted. Also animals that are about to become extinct can be produced.

Therapeutic cloning is helpful in producing organs and tissues that are healthy and that can replace the damaged organs or tissues by transplant methods. Scientists suggest that diseases like Alzheimer's or Parkinson's may be treated with this technology.

Unit 5

Nucleic Acids

5.1 BASIC CHEMISTRY OF NUCLEIC ACIDS

1. Cytosine

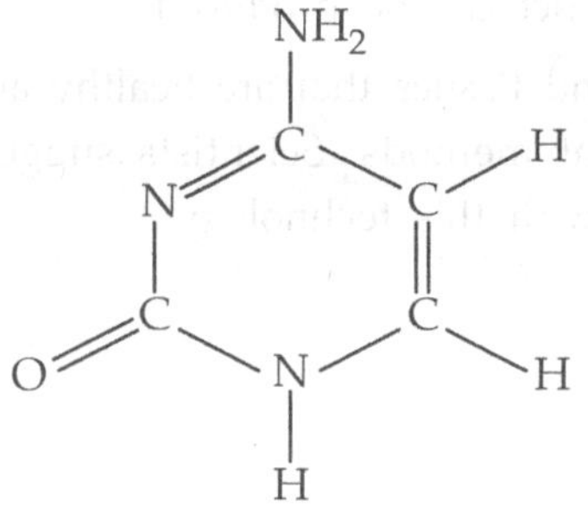

2. Uracil

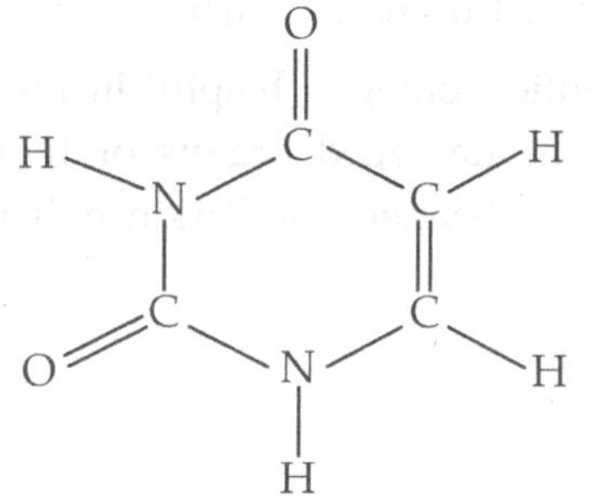

3. Thymine

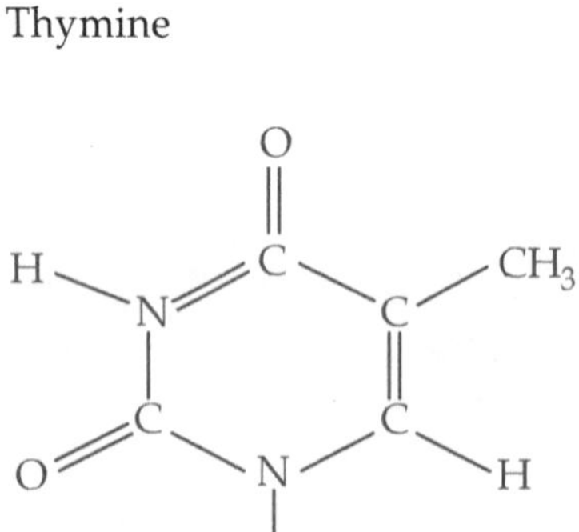

4. Adenine

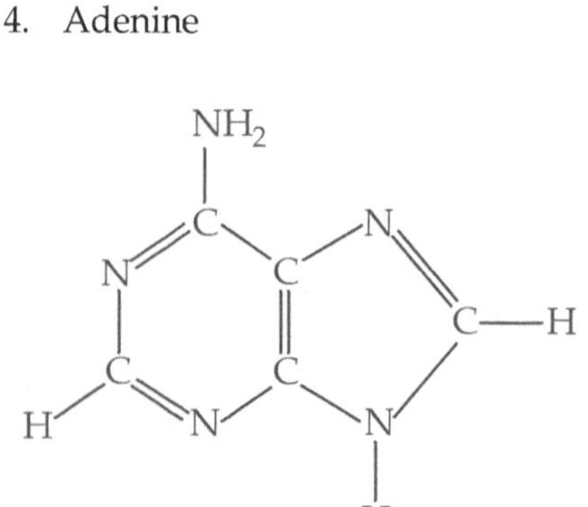

5. Guanine

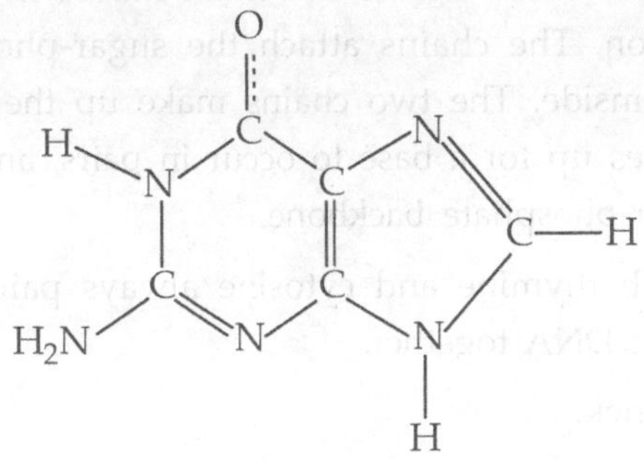

5.2 STRUCTURE OF DNA

DNA is a helix with a width of 2 nanometer . The bases are positioned one above the other at a distance of 0.34 nm in a form of steps. The steps make a full turn at a length of 3.4 nanometer. Therefore, at least ten base layers need to be located in one turn of DNA.

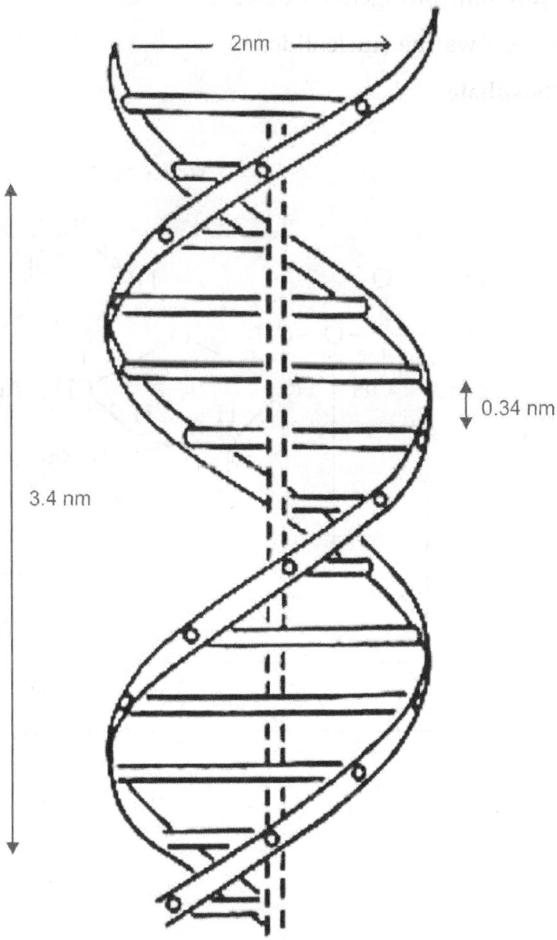

Fig. 5.1: The DNA Double Helix Structure

The two chains move in reverse directions, in which one of the chain moves in a 5′-3′ direction and the other moves in a 3′-5′ direction. The chains attach the sugar-phosphate component on the outside and the bases on the inside. The two chains make up the sugar-phosphate backbone. Each step of the ladder makes up for a base to occur in pairs, and each of these bases is connected to a base of the sugar-phosphate backbone.

In a DNA strand, adenine always pairs with thymine and cytosine always pairs with guanine. Hydrogen bonds hold the two strands of DNA together.

This model was proposed by Watson and Crick.

We just described the structure of the nitrogenous bases, now let us describe the nucleotides. There are three components present in each nucleotide:

1. A phosphate group
2. A central deoxyribose sugar
3. One of the four different nitrogenous bases

The following figures shows the nucleotides:

1. Adenosine Tri Phosphate

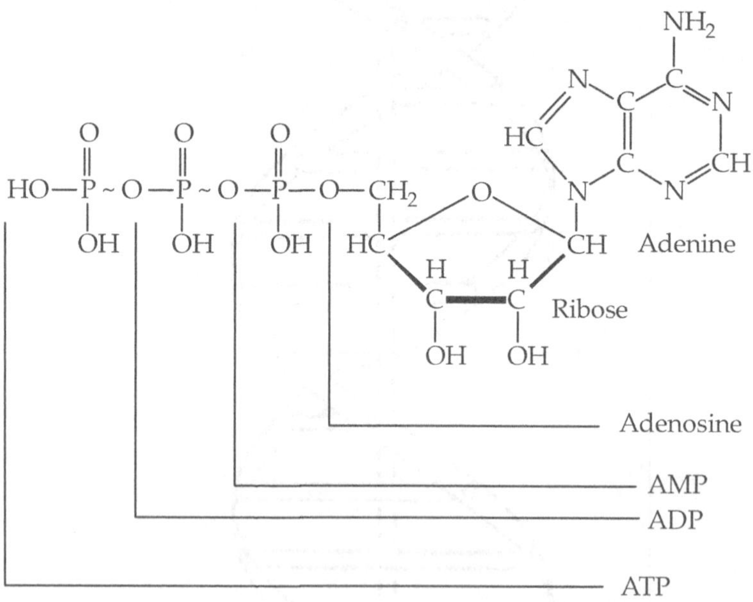

2. Guanine Mono phosphate

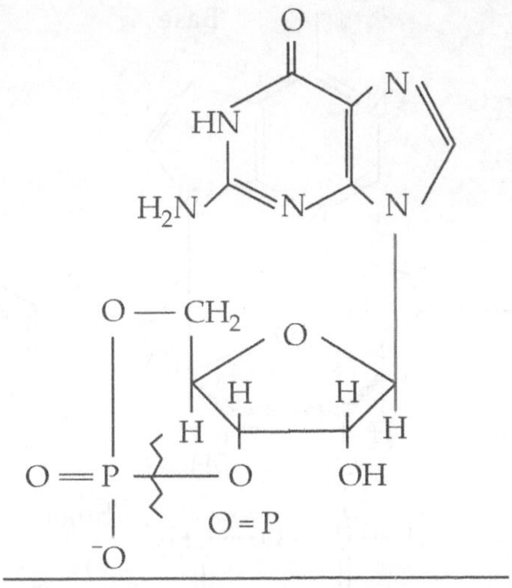

3. Cytosine Mono phosphate

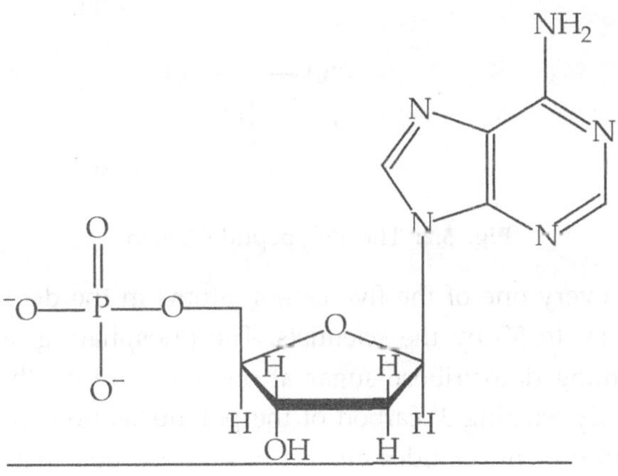

The nucleotide strings attach to each other and make a long polynucleotide, which when considered on a large scale becomes the chromosome. The nucleotides get attached to each other with the help of a phosphodiester bond. This bond joins the phosphate group of one nucleotide to the deoxyribose sugar group of the corresponding nucleotide. A link to the oxygen atom is called an Ester bond. Two ester bonds, each attached on either side of a phosphorous atom make up a phosphodiester bond.

The following figure shows the making of a phosphodiester bond. Nucleotides are added to the growing DNA only at their 3′ end.

Polypeptide chain showing phosphodiester bond

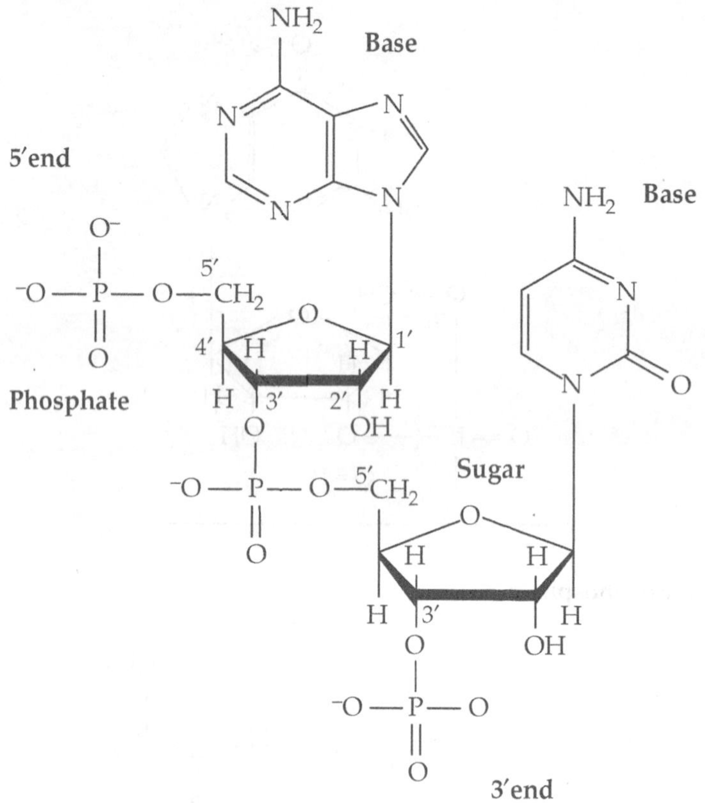

Fig. 5.2: The Polypeptide Chain

In the above figure, every one of the five carbon atoms in the deoxyribose sugar has been given a defined number (1′ to 5′) by the scientists. The phosphate groups fit themselves with the 5′ carbon of an incoming deoxyribose sugar at one end and on the other end they attach themselves with the already existing 3′ carbon of the polynucleotide chain. The result of which is that one end of the string of nucleotides always has a 5′ carbon that is free and at the other end of the nucleotide chain the 3′ carbon remains free.

5.3 BASE PAIRING

The term DNA sequence is used to tell the sequence of the nucleotides on a single strand arranged in any particular way. The supplementary strand that runs in the oppositely parallel direction in not in any random order but fixed according to the sequence given in the first strand. If a G is placed on a strand, a C will be placed corresponding to it on the other strand and vice versa. Similarly for any A on one strand, a T is found on its opposite strand and vice versa. The interaction between G's and C's and between A's and T's is balanced. This happens because Guanine is a two ringed structure and cannot be paired with another kind of two ringed structure (guanine and adenine are two ringed structures) because of the lack of space that exists between the two DNA strands. Similarly a single ringed structure (thymine) cannot pair with another single ringed structure, i.e., cytosine or another thymine. A and T form accurate spacing and balanced interaction and C and G also form yet another accurate spacing and balanced interaction. Chemical bonding is the most important aspect of pairing and apart from that spacing is another aspect which need consideration. A balanced and an energetically compatible condition arises when three hydrogen bonds are formed between guanine and cytosine and similarly when two hydrogen bonds are formed between adenine and thymine. This bonding alone is capable enough to hold the two strands of a DNA molecule together.

Even though the two strands of a DNA molecule are paired they are not in the same 5′/3′ direction. In fact, the two strands move in an antiparallel direction to one another, or in other words, if the 5′ end of a strand is pointing upwards and its 3′ end is pointing downwards, then in the opposite strand the 3′ end would be pointing upwards and the 5′ end would be pointing downwards. Taking an example, if a strand's nucleotide sequence is given by 5′-ACGTC-3′, then the other stand's corresponding sequence would be given by 3′-TGCAG-5′.

The following figure shows the bonding between the two pairs of nucleotides.

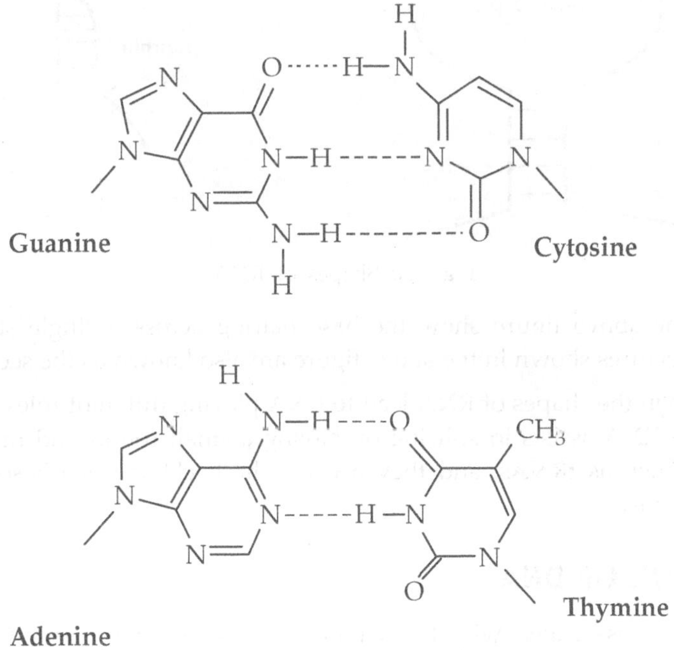

Fig. 5.3: Bonding Between Nucleotide Pairs

5.4 RNA STRUCTURE

The RNA mostly occurs as a single strand which is collapsed in an intricate fashion in such a way that base pairing occurs locally and also the interaction among hydrogen bonds exists. There are various ways in which an RNA may collapse and its complexities lead to different roles that the RNA plays in the cell.

RNA does not possess a long regular helical structure like the DNA molecule, instead round shapes are formed due to intra molecular hydrogen bondings.

RNA nucleotides are made up of ribose whereas DNA is made up of deoxyribose (a type of ribose that lacks one oxygen atom); and RNA has the base uracil rather than thymine that is present in DNA.

The ribose sugar in the nucleotides of RNA has its carbons labeled 1′ to 5′. A nitrogenous base is attached to the 1′ position which may be adenine, guanine, cytosine or uracil. A phosphate group fits between the 3′ position of one ribose and the 5′ position of its subsequent ribose. Hydrogen bonds are created between guanine plus cytosine and among uracil plus adenine and also among uracil plus guanine. Other types of bonding may exist to give the RNA different shapes and that is

1. Group of Adenine bases may attach to each other in the form of a bulge.
2. A hairpin loop or a stem-loop

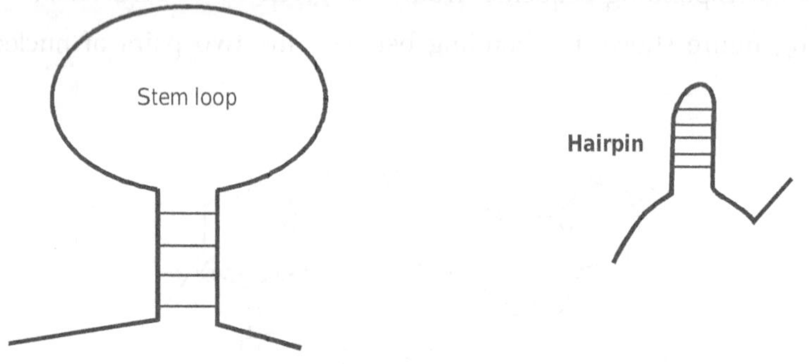

Fig. 5.4: Shapes of RNA

The lines in the above figure show the base pairing across a single strand of the RNA molecule. These structures shown in the above figure are also known as the secondary structures.

The differences in the shapes of RNA lead to RNA playing different roles on part of the cell. Taking an example, RNA assists in splicing on having a small round and nuclear shape, large RNAs are often known as tRNAs, and they become the backbone of ribosomes and assist in making of the proteins.

5.5 REPLICATION OF DNA

The term usually associated with DNA replication is that it is "Semi-Conservative". The meaning of Semi-Conservative is at the time when DNA is made into two or when it duplicates, the original DNA gives one strand and the other is made by the newly synthesized strand.

If the two strands of the original DNA molecule are separated then each of them provides the model for the formation of a new strand or a DNA molecule. We can find out how the new strand will look on the basis of the information we have in the original strand since the DNA sequence in one strand will choose the complementary bases on the new strand on the pretext that A will opt for T and C will opt for G. A conclusion drawn here is that one DNA molecule gives rise to two daughter DNA molecules which are exactly same as the original parent DNA molecule. Nonetheless each one of the descendant strands has one strand coming from the original DNA or the parent strand and the other is the newly formed complementary strand.

5.5.1 How It Replicates

The DNA starts replicating from a particular point and not just from anywhere in the strand. This point is also known as the origin of replication. As a result the two strands contribute in making a model for the creation of new DNA molecules, the two strands should separate. Therefore, in order to separate first the DNA should unwind and then break. To do so, there are special enzymes provided to do this. Helicase enzyme is required to unwind the helix and topoisomerase enzyme normally breaks and reseals the DNA strand. As soon as the model is organized and also the nitrogenous bases are all set on hand, the DNA strand will not be produced so quickly without the help of another enzyme known as the *primer*. The primer is nothing but a short chain of RNA formed on the DNA to begin the process since it is easy to add bases to a pre-existing chain of polynucleotides.

The primer is made with the RNA building blocks which are nothing but the four nucleotides (A, C, G, U) and another enzyme called the primase assists the primer.

The double stranded DNA is unwound up to a point forming a Y shaped structure known as the replication fork. The fork's point of divergence keeps moving as new strands grow on the fork.

The enzyme DNA polymerase polymerises the nucleotides in the 5′-3′ orientation in a continuous fashion and this is referred to as the leading strand. The other complimentary strand is known by the name of the lagging strand since it is synthesized in small pieces known as the *OKAZAKI* fragments. The fragments are formed with the help of the RNA primer and the primer is made with the help of the enzyme primase as we have already talked about. The enzyme DNA ligase serves in combining these fragments.

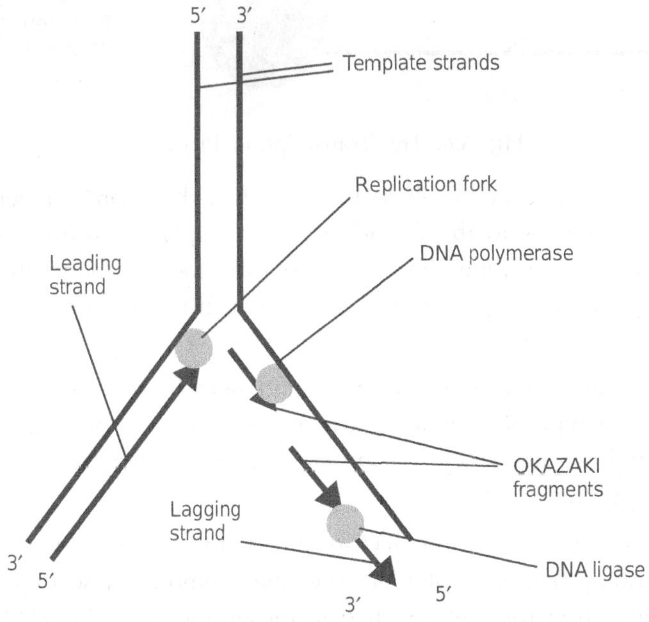

Fig. 5.5: The Replication Process

5.6 TRANSCRIPTION

DNA does not only replicate itself but furthermore constructs RNA. The method in which DNA is transcribed into RNA is known as Transcription. One of the DNA strands actually forms the RNA and the reaction is catalyzed by an enzyme known as RNA polymerase.

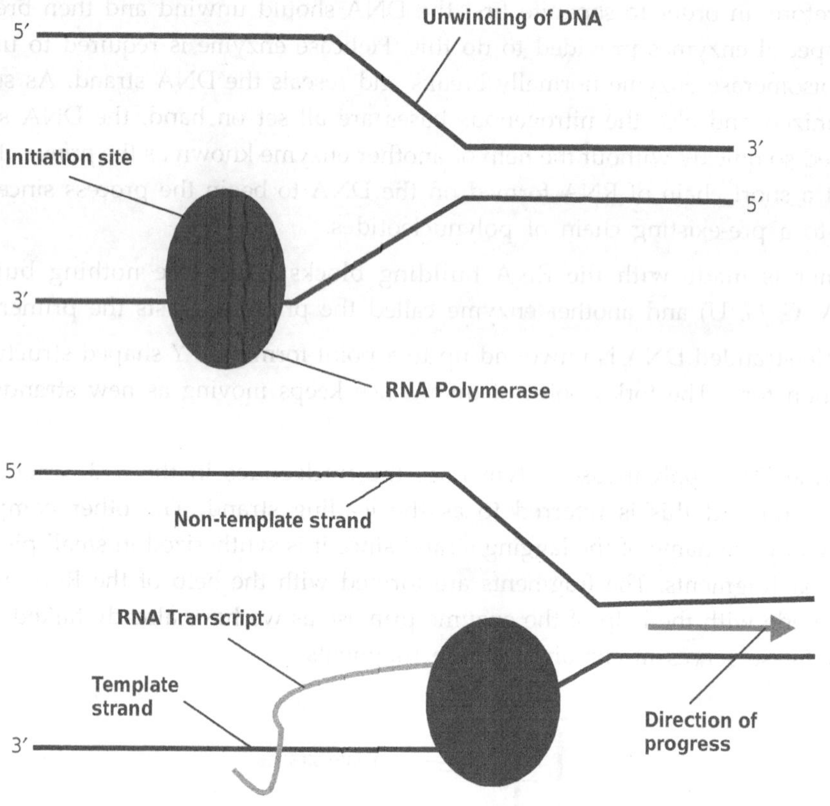

Fig. 5.6: The Transcription Process

The RNA polymerase attaches itself to the 3′ end of the promoter gene on the template strand of the DNA and moves to the 5′ end. The transcription process does not require any primers for initiation, in replication the same case was there, though the RNA polymerase needs a promoter sequence to be present in DNA to which it can attach with the help of other transcription factors.

The promoter sequence is around 10-35 bp and help in promoting transcription that is why it is called so. The type of promoters which are core promoters are present inside the promoter sequence and are necessary for the starting of transcription process. A commonly known core promoter is the TATA box.

Earlier we mentioned transcription factors, these transcription factors are only proteins which help in the binding of RNA polymerase to the promoter. It so happens that only after some specific transcription factors get attached to the promoter, is the RNA able to bind itself to the promoter. The unit consisting of RNA polymerase, promoter and the transcription factors is called the transcription initiation complex. After the first bond is formed, the promoter gets detached, then the RNA polymerase travels the template producing RNA using the base pairing complementary scheme. Only one DNA strand is used in the method. The transcription process comes to an end when the just made RNA strand forms a hairpin loop, followed by a number of Us which make it separate from the DNA template.

The DNA strand provides a template for the making of a number of RNA molecules, which are mainly of three types :

1. messenger RNA, known as mRNA
2. ribosomal RNA, known as rRNA
3. transfer RNA, known as tRNA

Messanger RNAs are RNA text of genes. They carry information from the DNA strand to the ribosomes which are the site for making of proteins. The transfer RNAs are untranslated RNA molecules that shift into the ribosome the building blocks of proteins or more commonly known as the amino acids. Finally the ribosomal RNAs are the untranslated components of ribosomes, made up of RNA and Protein compounds, rRNAs are occupied in grasping to the mRNA molecule and also catalyze some part of the translation process.

5.7 TRANSLATION

The making of mRNA into protein is where we get the genes to do their job in the cell. Similar to DNA, proteins are also strings of polymers built from an alphabet of chemically changeable elements. The protein's alphabet consists of a number of small molecules known as the amino acids. When comparing to the DNA molecule, the chemical sequence of a protein has physicochemical "content" as well as information content. Every one of the 20 amino acids more often found in proteins has a variable chemical character, determined by its side chain – a chemical grouping that differs from one amino acid to another amino acid. The chemical series of a protein is called its primary structure and when the protein folds into a compacted molecule, this is referred to as the secondary structure. The structures, both secondary and tertiary are mainly concerned with the function of the proteins. These secondary and tertiary structures bring close the far apart pieces of the chemical series to form what we call the functional sites. We assign the term genetic code to that part of the DNA that translates into protein. Nitrogenous bases come together in the form of 3 alphabets are more often referred to as a codon which make up an amino acid in the protein sequence. In DNA four nucleotides exist and the codon requires only three alphabets, so there are 64 possible combinations that can be formed from this. Therefore, in other words, 64 possible codons can exist and the protein alphabets are only 20 amino acids. So, in other words we use only 20 out of 64 combinations to assign to our amino acids. Some codons are redundant, some have been assigned special tasks of telling the cell to start the translation process or stop the translation process, these are called start codons and stop codons. The way the bases or the (nucleotides) are arranged in the RNA molecule determines the order in which the amino acids will be placed in the protein sequence.

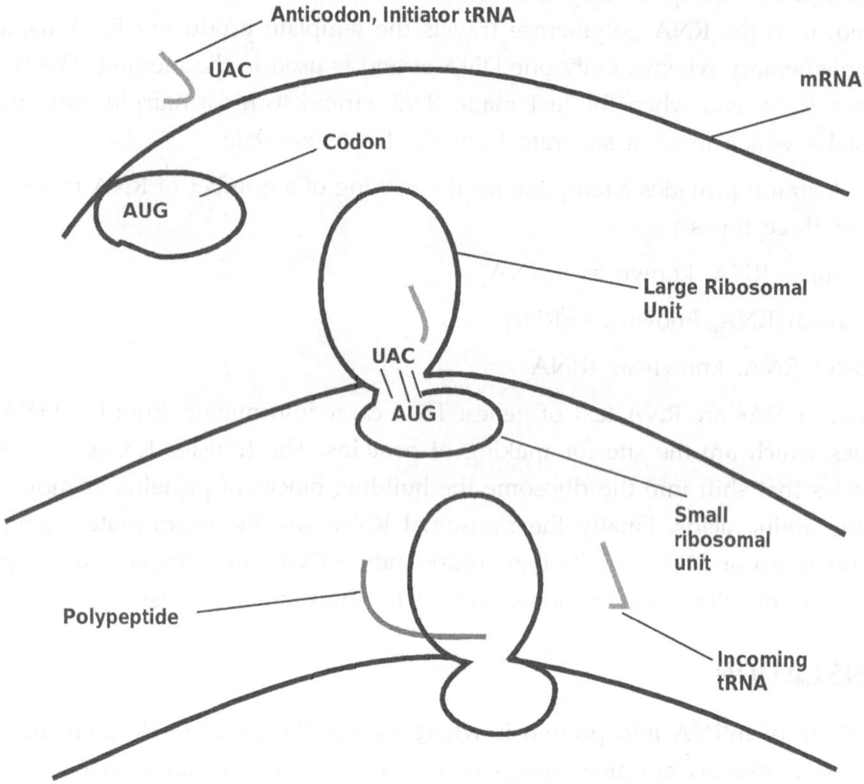

Fig. 5.7: The Translation Process

The tRNA charged with the amino acid supplies as an adaptor molecule to decode the information on the mRNA strand. This is made possible with the help of the tRNA interacting among the mRNA. The tRNA molecule is having a part which grasps the corresponding bases (anticodon) to the triplet code on the mRNA.

The proteins are released and later form a linear chain of polypeptides to do some useful job. The proteins which come out are put together with the help of peptide bonds between amino acids, where the COOH group of one may combine with the NH_2 group of another amino acid to form the CO-NH bond.

5.8 NUCLEIC ACIDS—INFORMATION STORAGE AND PROTEINS—INFORMATION EXPRESSION

The sequence of a particular gene may be extracted from a number of organisms and analyzed in order to find the function of the gene. The functions of a gene transform into some physical noticeable characteristics known as phenotype.

In earlier days, only one gene was put to test at a single instance, nowadays multiple experiments are carried out on a number of genes at the same time. Data from multiple genes of diverse organisms are incorporated to find out the function of the gene and the expression of the gene is studied to find out the phenotype of an organism. Further similarities among genes of varying organisms are evaluated.

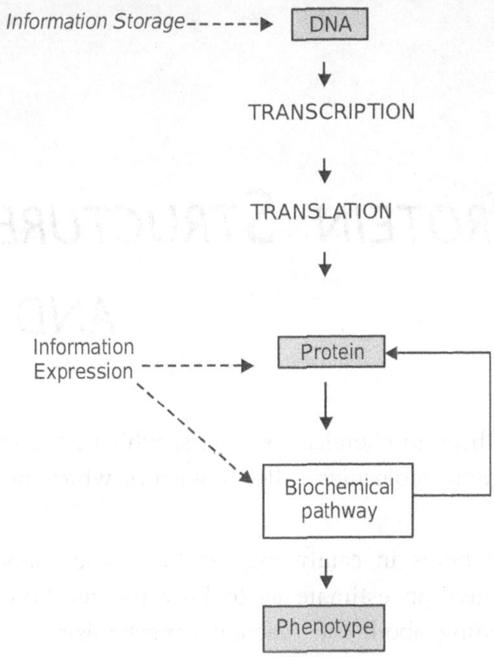

Fig. 5.8: Information Expression

The information storage or DNA holds the original message which corresponds to some sequence of adenine, cytosine, guanine and thiamine. Once the process of replication is over some information may be lost due to mutations, sickness or emission of rays and as a consequence the replicated copy of DNA may not be exactly like the first one. The information expression is nothing but the proteins where the nucleotide sequence is put to work and in the process of information flow from DNA to proteins, some information may be lost due to mutations etc.

Unit 6

PROTEIN STRUCTURE, FOLDING AND FUNCTION

The proteins function through chemical reactions, which are carried out with the help of amino acids and small organic molecules called cofactors which normally attach themselves to the proteins.

The protein's structure helps in catalyzing reactions, the shape of the protein and the placement of amino acids give an estimate as to how the reaction will take place, but this requires a lot of understanding about the reaction's mechanism of working.

Stability among proteins occurs due to hydrogen bonds and other interatomic forces that exist between them. Proteins themselves are drawn close to each other due to electrostatic forces. Proteins take shape due to interatomic interactions, and the shape also decides how intermolecular interactions will take place.

6.1 THE CHEMISTRY OF PROTEINS

The ID sequence of a protein is the basis for knowing the chemistry of a protein. Even though the 3D structure of a protein is complex, it has still arisen from the primary structure. In the linear chain of amino acids, some motifs were preserved and these motifs speak about the functions or structures of protein families. Thus the functional role of a protein is hidden in its structure.

The amino acids form a chain of polymer by forming a peptide bond between the basic amino group of one amino acid and the acidic carboxyl group of another amino acid. The result is a backbone structure. From the backbone the variable group of each amino acid projects out commonly known as the side chain. All the 20 amino acid side chains differ from each other in some way or the other.

The side chains can be categorized into many groups, some are large, some are small, some may have a positive charge, others may have negative charge, some are oily or hydrophobic (water-fearing), others may be hydrophilic (water-friendly). Some have big ring like structures, others may have straight carbon chains. Some are acidic others may be basic.

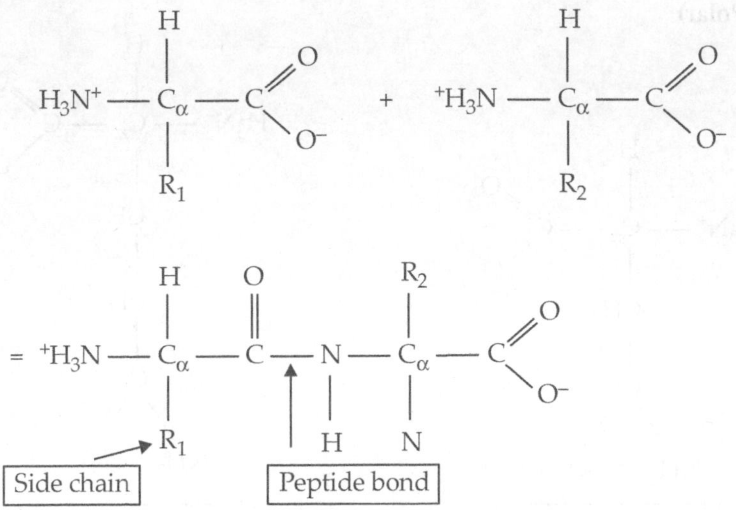

Fig. 6.1: The Amino Acid Chain

Amino acids are preserved in a sequence even after evolution because they perform some important functions such as :

1. To stabilize the protein structure
2. To form a specific binding site
3. To catalyze a reaction

The 20 Amino Acids along with their side chains

1. Small (Non-Polar)

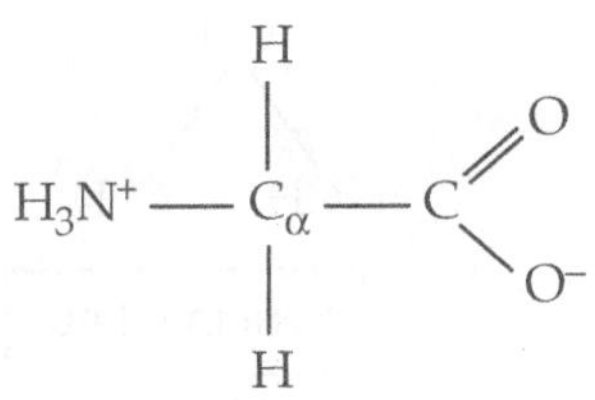

Glycine (Gly)

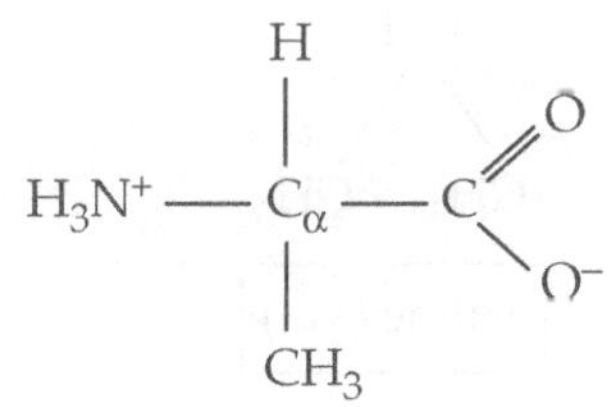

Alanine (Ala)

2. Small (Polar)

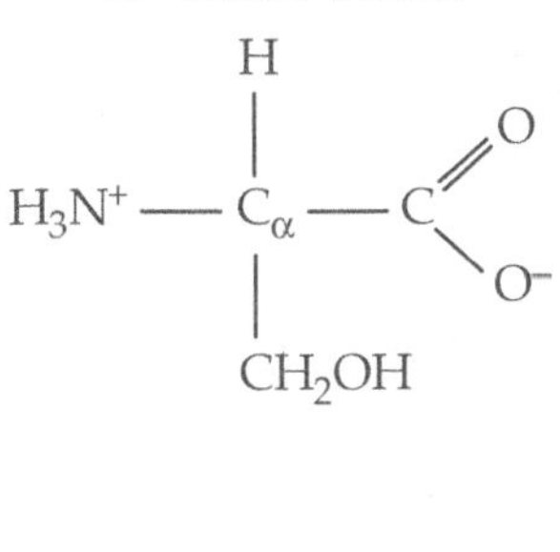

Serine (Ser)

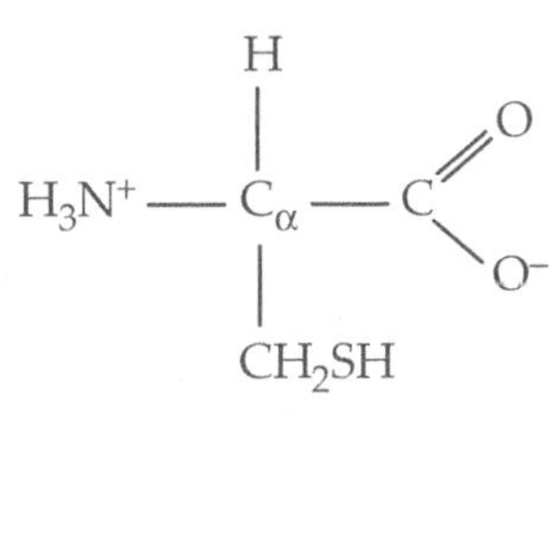

Cysteine (Cys)

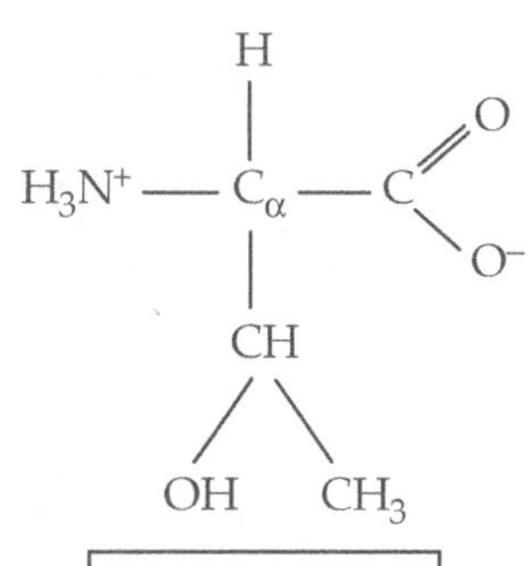

Threonine (Thr)

3. **Large (Polar)**

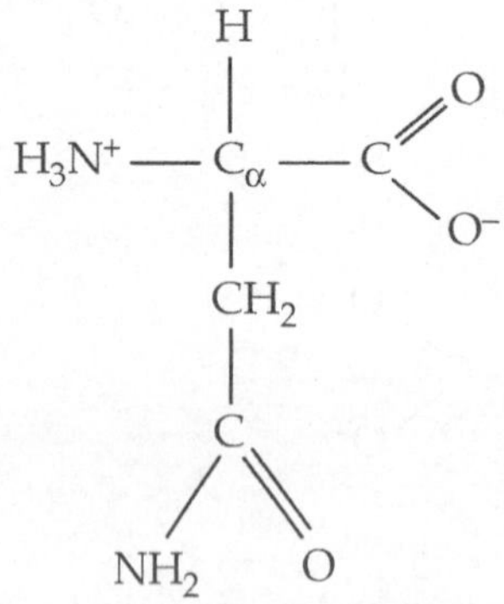

Asparagine (Asn)

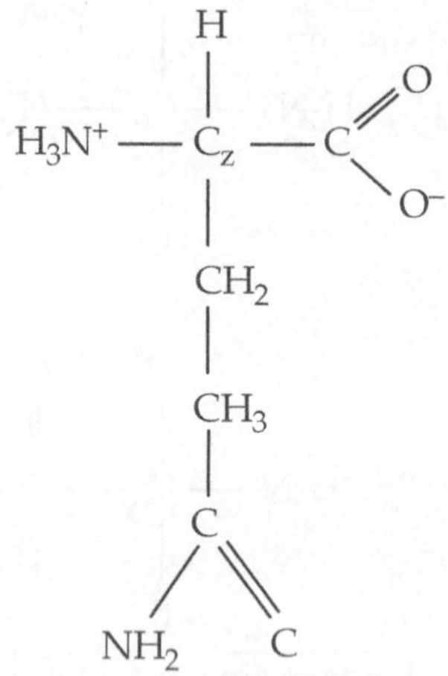

Glaxcine (Gln)

4. **Large Non-Polar**

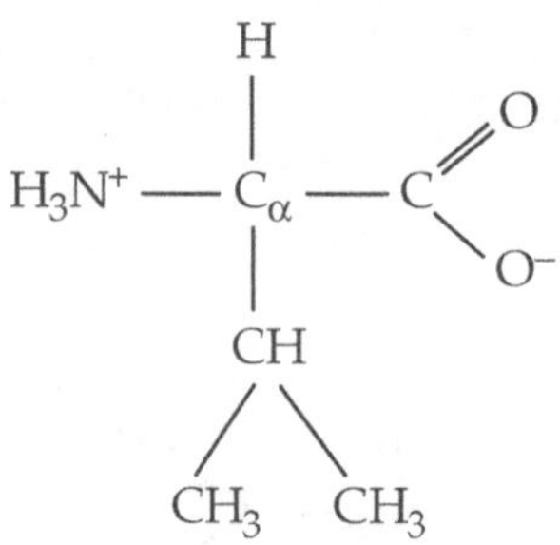

Valine (Val)

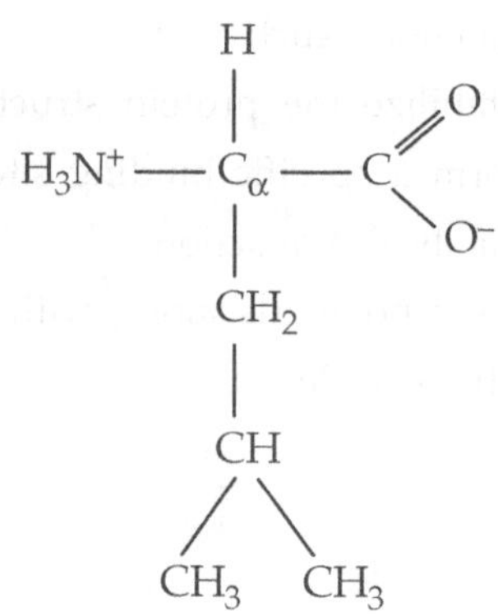

Leucine (Leu)

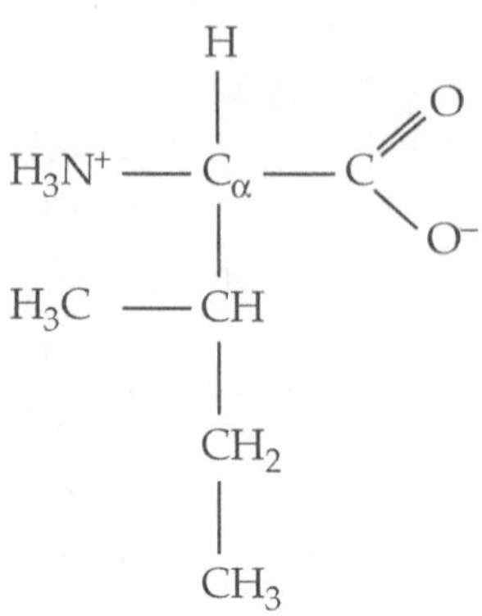

Isoleucine (Ile)

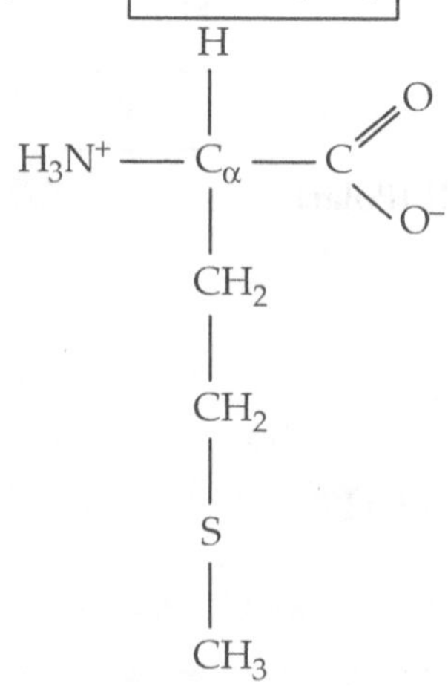

Methionine (Met)

5. Charged Basic

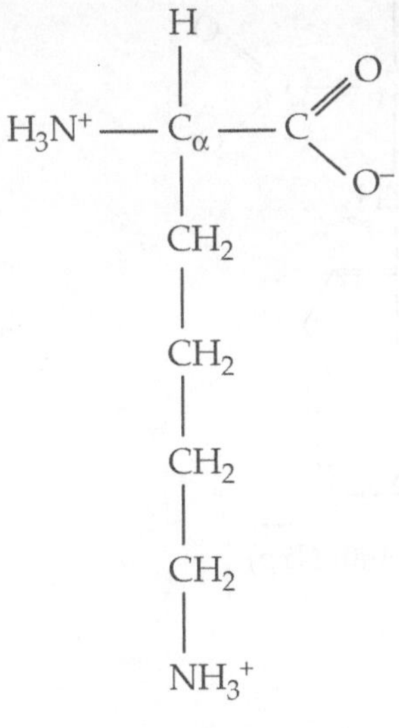

Lysine (Lys)

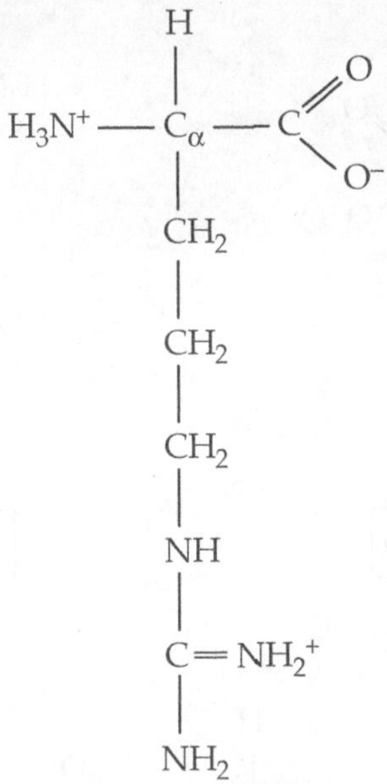

Arginine (Arg)

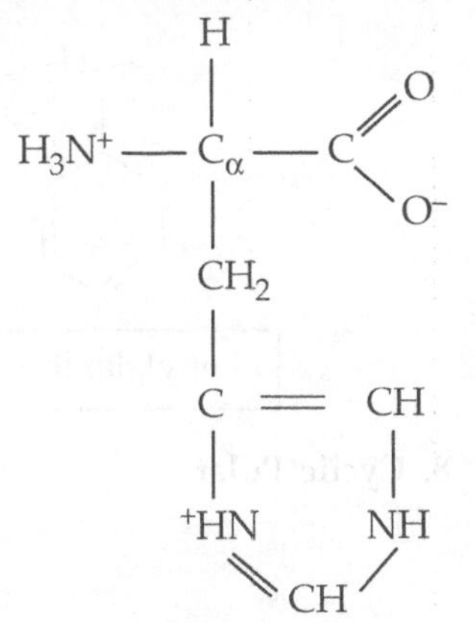

Histidine (His)

6. Charged Acidic

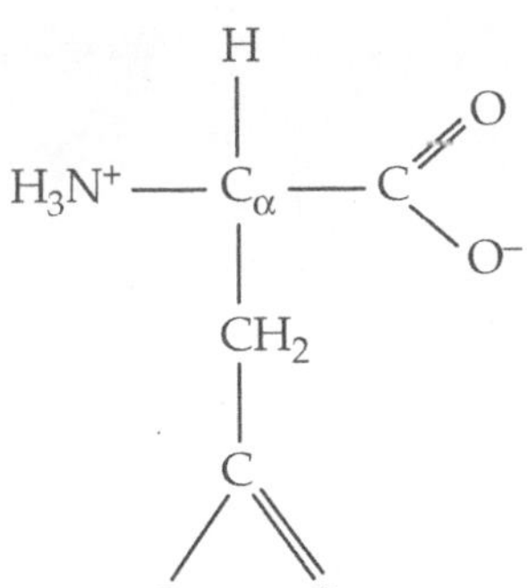

Asparatic Acid (Asp)

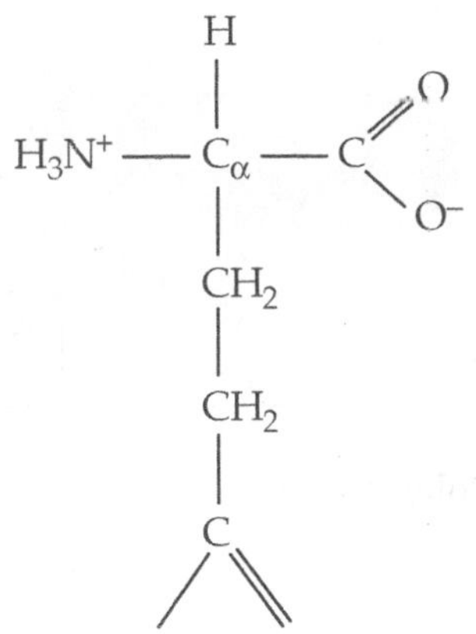

Glutamic Acid (Glu)

7. Cyclic, Non-Polar

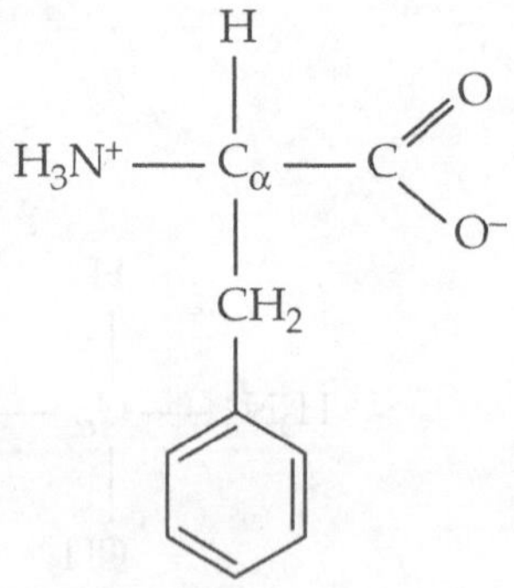

Phenylalanine (Phe)

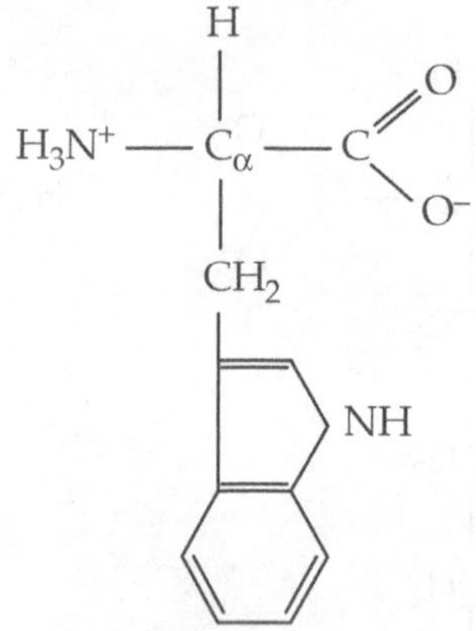

Tryptophan (Trp)

8. Cyclic Polar

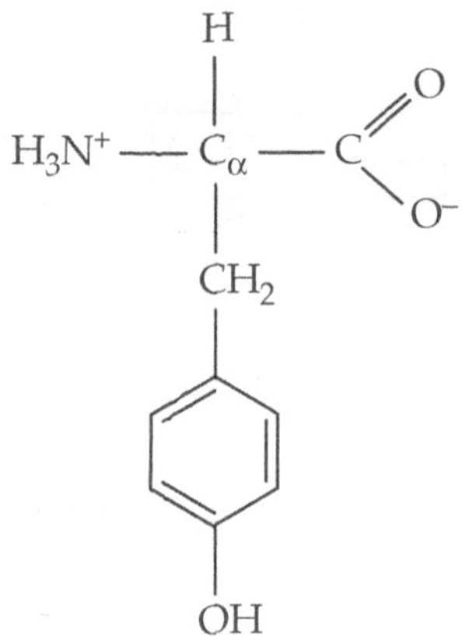

Tyrosine (Tyr)

---

9. Non-Polar Large

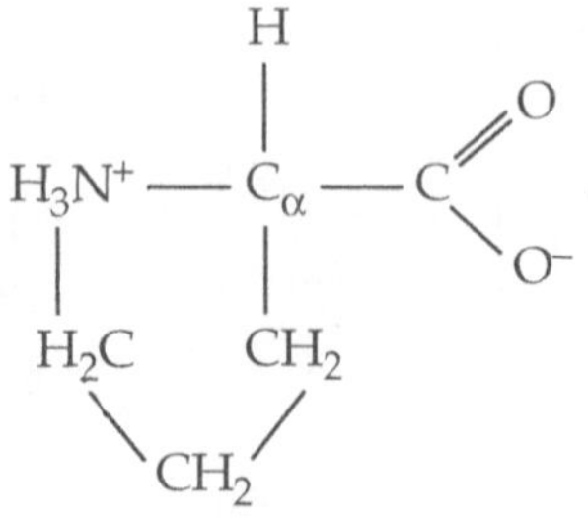

Proline (Pro)

The 20 Amino acids fall into more than one category. For example, an amino Acid side chain may be both non-polar and basic at the same time. The amino acid backbone is allowed to move in certain directions only, otherwise the atoms may bump into each other.

6.2 SECONDARY STRUCTURES OF PROTEINS

Pauling and Corey were the first to figure out the existence of alpha helices and beta sheets as a part of protein structures, from the known structures of amino acids.

Ramachandran was the first to explain the range of conformations (the flexibility to move) that were available to amino acids in a peptide chain. The meaning of a peptide chain conformation is the values that the dihedral angles may take in the protein backbone. The dihedral angles are the angles given by (1) the four atoms surrounding the N-C_α bond and (2) the four atoms surrounding the C_α – C bond. These angles are called the φ (phi) and ψ (psi) angles respectively. The peptide chain is not allowed to move around in any other third bond.

For any amino acid the values of φ and ψ are given and this represents the full description of the protein backbone. A Ramachandran map is simply a graph of φ versus ψ for a complete protein structure. A general Ramachandran map gives the permitted combinations of φ and ψ values for all the 19 non-Glycine amino acids in protein structures. There is a separate map for the Glycine amino acid. A protein structure may be evaluated by comparing its own Ramachandran map with the general Ramachandran map of permitted values of φ and ψ.

The following diagram shows a general Ramachandran map that shows the permitted combinations of φ and ψ values for amino acids in protein structures.

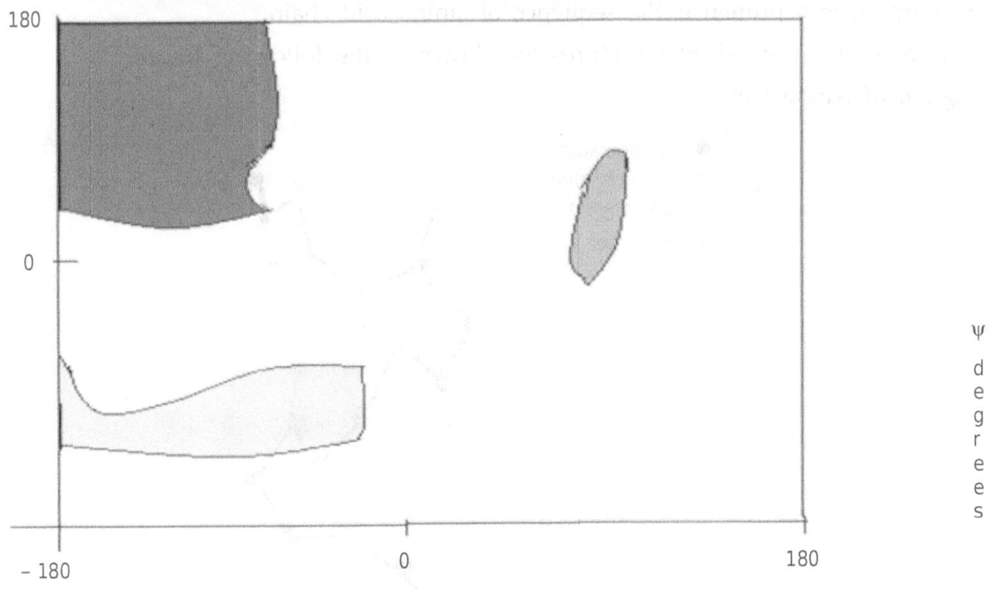

Fig. 6.2: The Ramachandran Map

The lower left half of the quadrant has a shaded region which shows the standard conformation of an amino acid in an alpha helix. The upper left half of the quadrant has a shaded region which shows the standard c φ (degrees) no acid in a beta sheet.

The following figure shows the φ and ψ angles in the backbone.

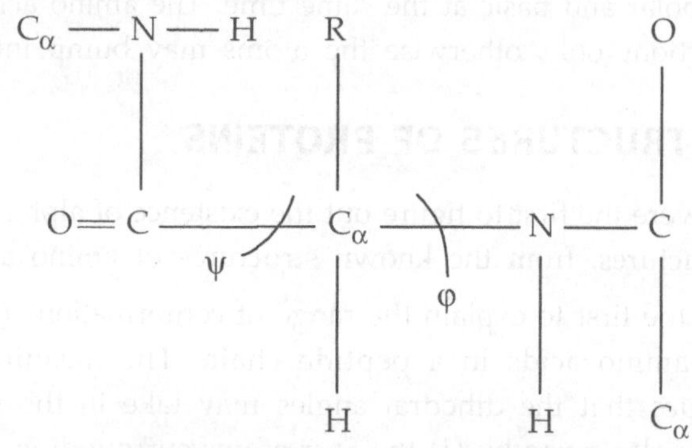

Fig. 6.3: The Amino Acid Backbone

The protein backbone moves along the bonds from the alpha carbon (alpha carbon is given the name to the carbon atom to which the side chain is attached) to the nitrogen atom and from the alpha carbon to the carboxyl carbon (carboxyl carbon is the carbon atom which has a double bond with the oxygen atom). These two bonds allow for circular rotation and are called the φ (phi) and ψ (psi) angles respectively.

From the Ramachandran map it is obvious that steric interactions are very vital determinants of the protein structure. Steric interactions immediately remove a large proportion of likely conformations for proteins and leave quite a small number of options for how a dense structure can form from a linear chain of amino acids.

The linear chain of amino acids is called the primary structure of a protein. The fundamental level of grouping in a protein is the sequence of amino acid chain.

Alpha helix and beta sheet structures are shown in the following figure:

Diagram of Alpha Helix

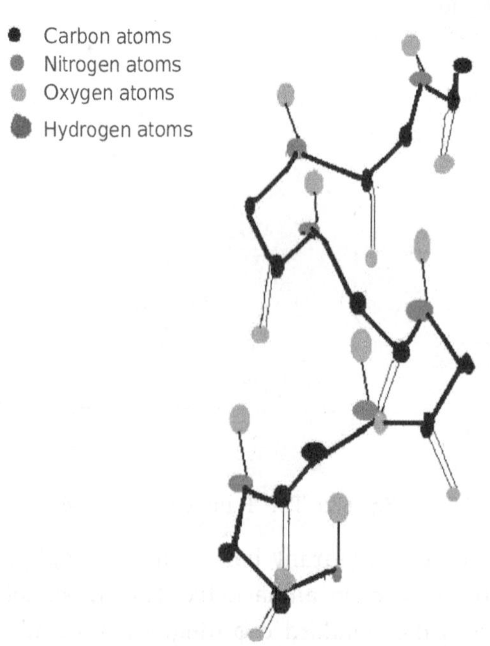

Fig. 6.4: The Alpha Helix

Diagram of Beta Sheet

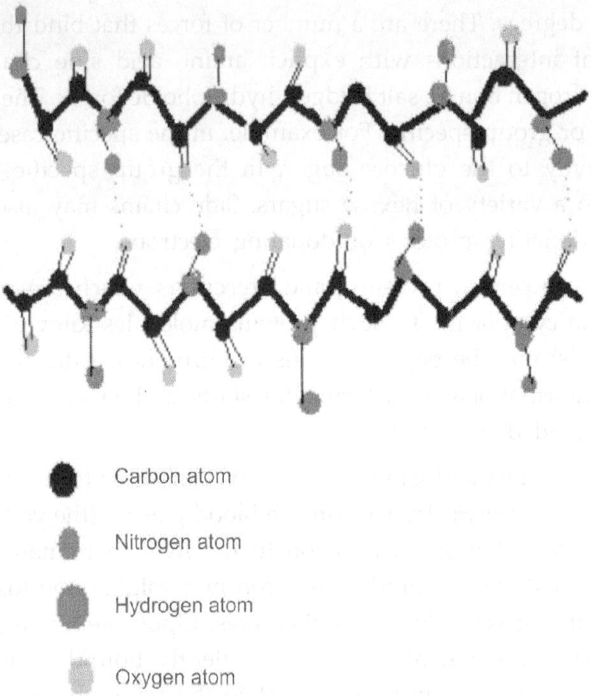

Fig. 6.5: The Beta Sheet

Alpha helix and Beta sheet structures are together referred to as secondary structures and are the next level of grouping. Normally when the amino acid chain is produced after the process of translation, the secondary structures are the first portions of the protein to fold.

Alpha helices are described by φ and ψ angles of approximately – 60° and show a spring like structure with 3.6 amino acids per complete 360° turn. Beta strands are described by φ angle of approximately – 135° and ψ angle of approximately 135° which gives it an almost linear backbone shape.

6.3 PROTEIN FUNCTION

Proteins are known for doing the majority of the work in both prokaryotic and eukaryotic cells. The functions of proteins differ largely. For example, collagen is a structural protein which gives rigidity and support to bones and connective tissues. Proteins also act as catalysts in the form of enzymes, for example, pepsin is a protein that assists in breaking down the food and also metabolizes it. Proteins also carry atoms and small molecules to different parts of the organism, for example, hemoglobin is a protein which carries oxygen to different parts of the body. Proteins also carry information to other cells, for example, insulin carries amino acids to other cells. Proteins also absorb elements from external environment, for example, Rhodopsin absorbs light to enable vision. And there are plenty more functions.

Let us describe the functions of proteins in a little more detail:

1. *Enzymes:* All enzymes are proteins. These can increase the speed of biochemical reactions by a number of degrees. There are a number of forces that bind the substrate and these are non-covalent interactions with explicit amino acid side chains such as van der Waals forces, hydrogen bonds, salt bridges, hydrophobic forces. The binding of substrates may be specific or group specific. For example, in the specific case, the glucose oxidase enzyme binds only to the glucose sugar. In the group specific case, the hexokinase enzyme binds to a variety of hexose sugars. Side chains may also help in the process of catalysis by donating protons or donating electrons.
2. *Signaling:* There are certain proteins called receptors which are present in the plasma membrane or the cytoplasm. To such proteins molecules known as ligands bind. The molecules (ligands) may be peptide chains or hormones. After the binding the protein undergoes a conformational transform and starts a chemical reaction within the cell. Normally the ligand does not change.
3. *Transport and Storage:* Hemoglobin transports oxygen in the form of red blood cells to the whole body of the organism. Transferrin is a blood plasma (the yellow liquid component of the blood) protein that shifts iron ion to the liver in humans. Ferritin, a globular protein (which is soluble in water) stores iron in a soluble non-toxic state in both type of cells namely, the prokaryotes and eukaryotes. Lipoprotein is a gathering of proteins and lipids which are covalently or non-covalently bound to proteins. The task of lipoproteins is to carry fats and cholesterol to the whole body of the organism with the help of blood in which they might usually dissolve.
4. *Structure and Movement:* Collagen is the main protein of connective tissue (a tissue is a group of cells that perform alike functions). The task of collagen is to give cells structure from the outside. Collagen may also be found in skin and bones of the organism. They give extra support to the skin where needed. Keratin is a fibrous structural protein which is hard and insoluble. The skin, hooves, feathers, fur, nails, hair, horns, beaks, wool, claws and scales are made up from keratin. Actin is a globular, highly conserved protein found in all eukaryotic cells. Actin is responsible for (1) carrying out the process of muscle contraction, (2) quick movement of the cell, (3) cell division, (4) vesicle and organelle movement, (5) helps the cell to receive information from and respond to the environment in order to develop move along actin filaments while hydrolyzing (splitting of water molecule into hydrogen and hydroxide ion) ATP. They mainly help in mechanical movement of the cell in reaction to some input. The energy required to do mechanical work is released by hydrolyzing ATP, the cell, repair the tissues and make the cell immune and finally (6) to make and maintain cell shape. Myosins are family of motor proteins that
5. *Nutrition:* Proteins are present in our food. Casein is a protein that makes up almost 80% of proteins in cow's milk and cheese. Casein is a kind of phosphoprotein (phosphoprotein is a protein that binds to phosphoric acid). Casein consists of a number of proline peptides. Ovalbumin is a protein that makes up about 60%-65% of proteins in the white part of an egg. These proteins provide amino acids for the growth of developing

offspring. The seeds of many plants contain proteins which can provide a good protein diet, apart from that seed proteins give nutrition to the embryos during the process of plant germination.

6. *Immunity:* Immunoglobins are gamma globular proteins which are also commonly known as antiobodies are present in the blood and fluids of an organism. The antibodies attack the viruses and bacteria that come into body. An antigen is also a protein. An antigen when introduced in the body causes the immune system to produce antibodies.
7. *Regulation:* Genetic regulatory proteins bind to segments of DNA and bring about gene regulation. In gene regulation, transcription factors are used which are proteins themselves and the process is called transcription.

6.4 PROTEIN STRUCTURE

The linear chain of amino acids has a free amino group (NH) at one end and a carboxylic acid (COOH) group at the other end. The numbering of amino acids starts from the amino terminal and moves towards the carboxylic terminal.

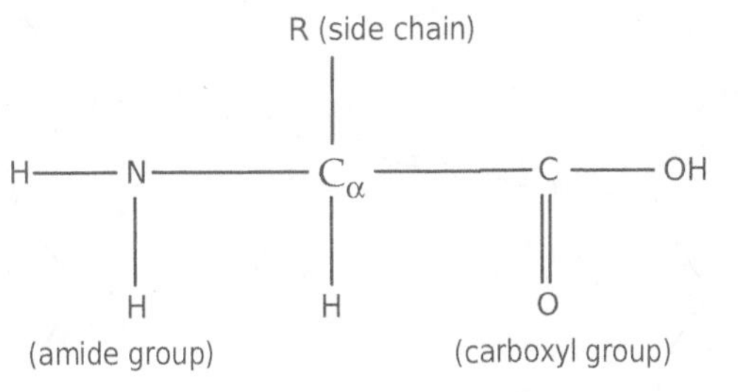

Fig. 6.6: The Linear Chain of Amino Acids

After the process of translation a protein does not remain in its primary structure, i.e a linear chain of amino acids but falls down and folds into a complicated spherical shape. The arrangement of amino acids on the linear chain mainly decides the shape that a protein will take after it folds. The exclusive structure that a protein takes after the fold is called the native structure of the protein.

The native structures of proteins have distinct properties which make them responsible for carrying out specific tasks in a living organism. For example, hydrophobic amino acids have side chains made up of by and large or completely carbon and hydrogen atoms and they are not likely to develop hydrogen bonds with water. On the other hand the polar amino acids which have side chains made up of oxygen or nitrogen atoms easily develop hydrogen bonds with water. The charged amino acids carry a positive or negative charge. The arrangement of amino acid sequence in the linear chain is responsible for depicting the secondary structure and ultimately the tertiary structure of the protein.

The properties of amino acids give their proteins its shape and function. Or we can say that the sequence of amino acids in a protein are responsible for giving it a 3D structure, its chemical and physical characteristics and finally the function it plays in the body.

To foresee the three dimensional structure that will be formed from the primary structure, we would need to first find out the sections of the backbone that are expected to form alpha helices and beta turns. A beta strand forms an anti-parallel structure when it takes a U turn or runs in the opposite direction.

As it must be clear by now that secondary structures are of two types, alpha helices and beta sheets. The beta strands gather into either a parallel beta sheet or an anti-parallel beta sheet. In a parallel beta sheet, the two strands move in the same direction. In an anti-parallel beta sheet neighboring strands move in reverse directions as you go down the protein backbone from amino group to carboxyl group.

The following diagram shows the parallel and anti-parallel beta sheets. In an anti-parallel sheet the structure is more energetically stable due to hydrogen bonding than a parallel sheet.

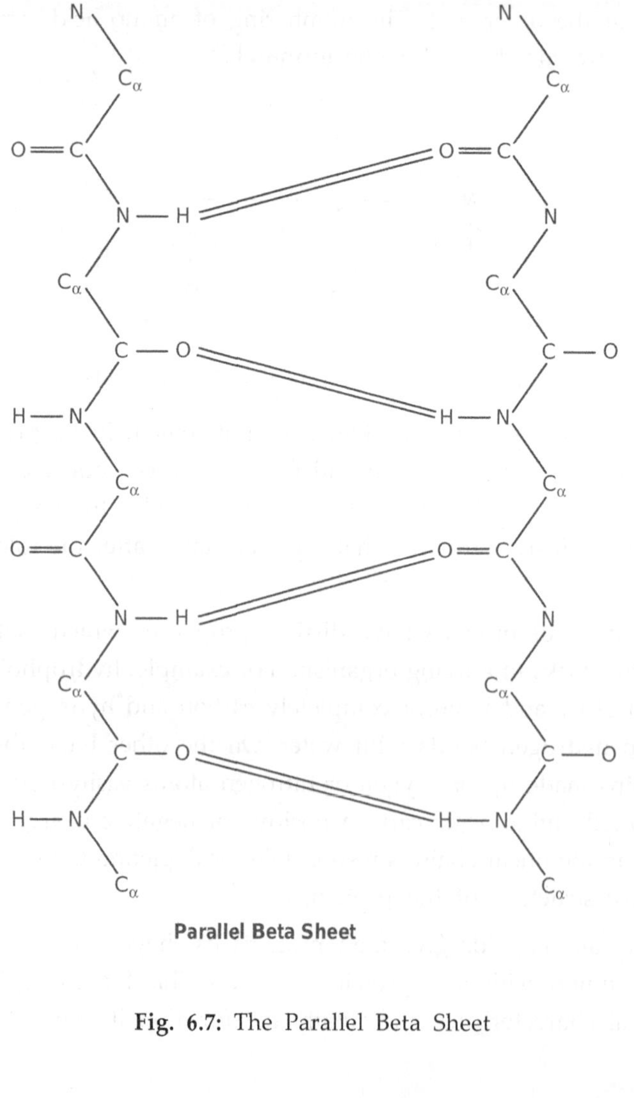

Parallel Beta Sheet

Fig. 6.7: The Parallel Beta Sheet

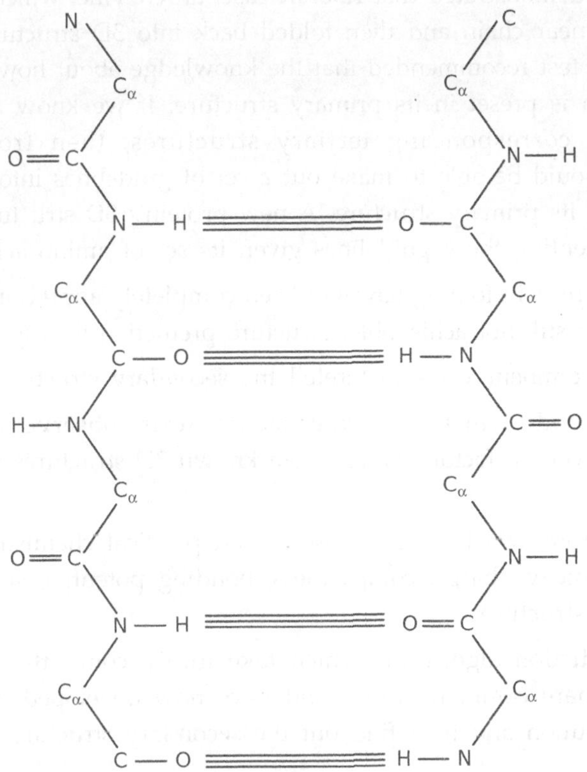

Fig. 6.8: The Anti-parallel Beta Sheet

The secondary structure packs itself to give a three dimensional structure known as the tertiary structure of the protein. An active enzyme is usually formed by two or more protein chains and is combined into a single large complex. In this the proteins interact with each other and is known as quaternary structure.

6.5 PROTEIN FOLDING

The foretelling of secondary, tertiary and quaternary structures of linear chains based on their primary structure is what the protein folding problem is all about. Research is going on into finding out what forces are responsible for the folding of proteins. It is already proven that a number of forces such as hydrogen bonding, electrostatic nature of forces and van der Waals forces play an significant role in the folding of proteins. The making of covalent bonds between cysteine residues play an important role in deciding protein folds. The problem is further complicated by the work of particular group of proteins known as *chaperons*, which change the protein structures in random but keyways.

The basic question that comes to our mind while dealing with the protein folding problem is that if a linear chain of amino acids is given to us then how does this primary structure decide the final 3D fold that it is going to adopt.

In 1961, Anfinsen demonstrated that ribonuclease, an enzyme, which is actually a protein was unfolded into a linear chain and then folded back into 3D structure without loss of its biological activity. This test recommended that the knowledge about how the protein will take its native conformation is preset in its primary structure. If we know a number of primary structures and their corresponding tertiary structures, then from analyzing those correspondences we should be able to make out a set of guidelines into what 3D structure a protein will take given its primary structure. A new protein's 3D structure could therefore be determined by implementing those guidelines given its set of amino acid chain.

The guidelines for protein folding have not been completely appreciated and predicting the structure of a protein is still not achievable. Structure prediction is only about 50-60% certain.

There are three prominent ways to foretell the secondary structure of a protein.

1. There are methods which use experiments, tests, observations, experience and calculations on certain factors gained from known 3D structures to find out secondary structures.
2. There are methods which base themselves on physical chemistry properties such as fold, hydrophobicity, charge, compactness, bonding potential, size etc., to determine the secondary structure
3. There are prediction algorithms which take into account the known structures of proteins that share common origins and have now developed differently due to the process of evolution and then find out the secondary structures.

6.5.1 Chaperons

Chaperons are molecules of proteins whose primary job is to help the proteins to fold correctly within the cell. They are also responsible for stopping aggregation among proteins. They generally obstruct erroneous folding pathways that develop into nonfunctional products. There are many types of chaperons and have a variety of functions. Assembly chaperons are one such type which help in the assembly of nucleosomes from folded histones and DNA. In the nucleus they convert folded subunits into a polymer that consists of two, three, or four monomers. Some chaperones help in folding the newly synthesized proteins as they are pushed out from the large ribosomal unit during the process of translation and also prevent them from aggregating into structures.

Other types of chaperones are involved in transport across membranes, such as in eukaryotes they transport in the mitochondria and in the endoplasmic reticulum (ER). New functions for chaperones have been discovered, for example, assisting protein degradation, and responding to diseases linked to protein aggregation.

6.5.2 Hydrophobicity

The compressed spherical shape that the proteins take on is due to a keyforce involved in protein folding known as the *hydrophobic effect*. This happens because the proteins try to maintain a minimum number of hydrophobic side chains lay open to water. This is the main rule associated in the folding process of proteins.

The native conformation that most proteins adopt is having a hydrophobic core. In the *hydrophobic core,* all the hydrophobic residues are drawn towards the centre leaving space for the charged and polar residues to occupy the solvent open surface. The process in which the hydrophobic residues are set apart from the solvent is commonly known as the *hydrophobic collapse.* The membranes are normally made up of hydrophobic amino acids which have carbon and hydrogen as their atoms. These molecules are guarded from water molecules.

The Hydrophobic Collapse

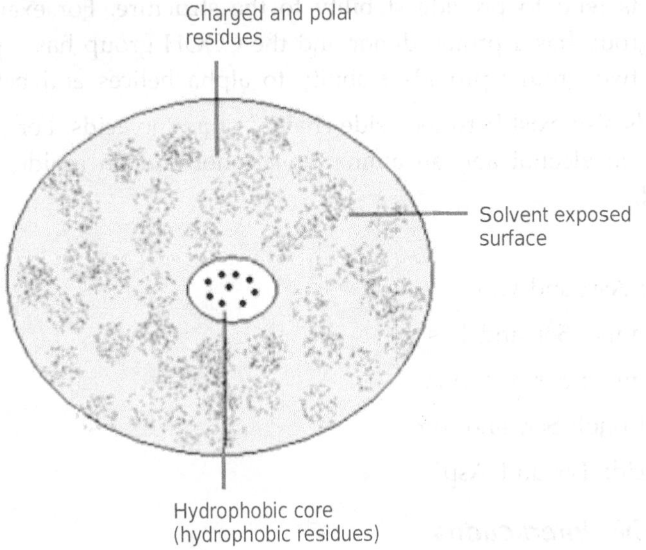

Fig. 6.9: Hydrophobicity

6.5.3 Interatomic Forces and Protein Structure

Interatomic interactions play a key role in deciding the 3D shape that the protein is going to take. Its chemical properties are also determined by interatomic interactions.

6.5.3.1 Covalent Interactions

Covalent Interactions are very short range mighty forces varying between 1 to 1.5 Angstroms. Due to these, the atoms are closely bound into a molecule. Sharing of electrons takes place in covalent bonds. The covalent bonds are also responsible for altering the nature of atoms in the molecule. In covalent bonds the atoms do not exist as separate elements but come together to form one molecule. A prominent example of covalent bonding is the peptide bond between amino acids of the protein backbone. Also in the individual side chains of amino acids, the atoms are bonded covalently. The covalent bonds are responsible for determining the length between the atoms of a protein structure. A covalent bond distance is based on the number of electrons shared between atoms, the type of atoms and the size of atoms. Generally shorter stronger bonds appear as a result of more number of electrons being shared.

6.5.3.2 Hydrogen Bonds

A short strong bond exists when two polar groups come together and is known as the hydrogen bond. The two polar groups are of particular type, one is a proton donor, the other is a proton acceptor. The proton donor is positively charged and the proton acceptor is negatively charged and the bond between them is said to be polarized. The proton donor, for example, is a proton or a hydrogen atom which is attracted towards a proton acceptor, for example, an oxygen atom which is strongly electronegative.

Hydrogen bonds tend to provide stability to the structure. For example, in the protein backbone, the NH group has a proton donor and the COOH group has a proton acceptor. The forces between the two groups provide stability to alpha helices and beta sheets.

Hydrogen bonds also exist between "side chains" of amino acids. For example, it may exist between two acids, an alcohol and an amine, an alcohol and an amide, two alcohols, or an alcohol and an acid.

Examples:

Acid and Acid: Asp and Glu

Alcohol and Amine: Ser and Lys

Alcohol and Amide: Ser and Asn

Alcohol and Alcohol: Ser, and Tyr.

Alcohol and Acid: Tyr and Asp

6.5.3.3 Hydrophobic Interactions

The non-polar side chains have hydrophobic interactions and therefore are responsible for providing stability to tertiary structures in proteins. The non-polar groups withdraw from water and from other polar groups and are drawn to other non-polar groups like themselves. Hydrocarbon alkyl groups such as Ala, Val, Leu, and Ile are drawn towards each other and other benzene (aromatic) rings such as Phe and Tyr are drawn towards each other.

6.5.3.4 Charge-Charge, Charge-Dipole and Dipole-Dipole Interactions

Other forces also exist in the protein conformations and they are not definite. The atoms in such forces do not share electrons as is the case in covalent bonds. The nature of the atoms also remains unchanged. These are the forces that appear in polar and charged atoms. The charged atoms are the ones that either have a positive charge or a negative charge. Such charged atoms are called ions. An ion has a positive charge if there is a deficiency of electron and it is said to have a negative charge if there is a surplus of an electron.

Charge-charge interactions exist among the positively and negatively charged side chains of acidic or basic amino acids. In the case of proteins, these charge-charge interactions are also called *salt-bridges* and they are responsible for providing major stability to the proteins.

Dipole-dipole interactions are attractive forces between the positive end of one polar molecule and the negative end of another polar molecule. They are much weaker than ionic forces or covalent bonds and have a keyrole to play only if the concerned molecules are close together (touching or almost touching). The dipole of a molecule is actually a vector which gives the length of the polarization along a bond. In proteins, the amino acid side chains as well as the protein backbone shows a strong dipolar force which helps in further stabilizing the protein structure.

6.5.3.5 Van der Waals Forces

The van der Waals force is a temporary, weak electrical pull of one atom for another. Van der Waals forces arise because every atom has an electron cloud that can vary, giving a momentary electric dipole. The short lived dipole in one atom can cause a complementary dipole in a different atom, given the situation that the two atoms are relatively near to each other. These temporary, complementary dipoles supply a weak electrostatic force, the van der Waals force. This is a fairy long range force similar to gravity. They differ from covalent and ionic bonding in that they are caused by correlations in the varying polarizations of nearby particles.

Van der Waals force, even though temporary and weak, can supply a keyrole to protein structure because of its infinite number. Most atoms of a protein are packed quite close to others to be involved in temporary van der Waals forces.

6.5.3.6 Repulsive Forces

Steric interactions or repulsive forces arise when two atoms come close together and their electrons start repelling each other. These are very short range forces, obviously because they only arise when atoms start coming close to each other. Each atom has its own circular limit, also known as the van der Waals radius. One atom's circular limit cannot cross another atom's circular limit, because if they start crossing, the repulsive forces begin to arise. Normally atoms try to stay away from each other's private space.

6.5.3.7 Strength among Interatomic Attractions

Atoms are attracted to each other at a certain distance, if the distance is reduced they start repelling each other and when the distance is increased, the attraction also subsides. When the stability of a protein is concerned, then these interatomic forces play an important role in the stability and interatomic forces depend on the strength of the attraction between atoms and at what distance they are placed.

6.6 NUCLEIC ACID PROTEIN INTERACTION

The nucleic acid protein interactions have been seen in almost a set of 250 DNA Protein complexes. The way nucleic acids interact with proteins in eukaryotic cells is different from the way they interact in prokaryotic cells. Let us first describe the prokaryotic cell interaction and then the eukaryotic cell interaction.

6.6.1 Prokaryotic Cell Interaction

A commonly known prokaryotic cell is that of the *E. coli.* In an *E. coli* chromosome, the DNA length is almost equal to 4.6 million base pairs and has a single closed circular strand which is crowded into a compact area known as the nucleoid. The nucleoid contains DNA in a very high concentration along with the proteins which are connected with DNA such as polymerases etc. Here is an example of what the nucleoid of *E. coli* may look like:

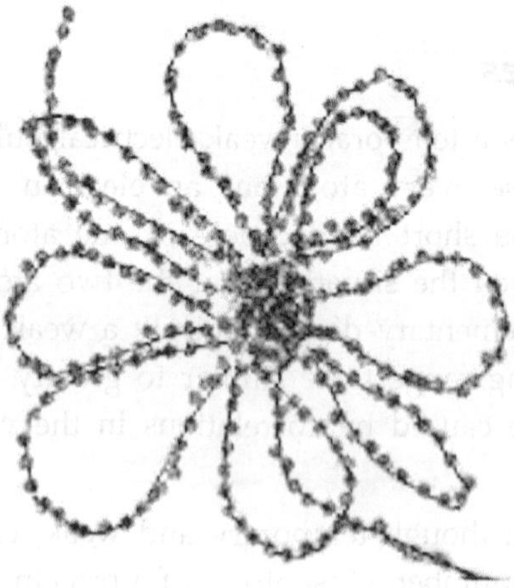

Fig. 6.10: Nucleoid of *E. coli*

The DNA consists of 50-100 loops which are bound at the base to a protein membrane core which restricts the rotation of the DNA molecule.

6.6.2 Eukaryotic Cell Interaction

In a eukaryotic cell, the length of the DNA depends on the species but in most cases is thousands of times greater than prokaryotic cell. In eukaryotes, the DNA is identified by chromosomes which are present inside the nucleus. The chromatin is a nucleoprotein complex which contains more than 50% of proteins and it changes its structure during cell division. The chromosomes are made from chromatin at some stage of cell division.

In simple terms, DNA wraps around nucleosomes, a chromatin fiber consists of an array of nucleosomes, the chromatin turns further into chromosomes.

The proteins found in the chromatin consist of histones, which is a family of proteins. All histone proteins carry a positive charge and around 20-30% of their sequences consist of basic amino acids lysine and arginine. The histones therefore bind well to negatively charged DNA when they form chromatin. All histone classes closely resemble in their tertiary structure. Histone H1 is estimated to be present in half of the chromatin structure.

The chromatin structure is described below:

A basic unit of chromatin structure is called the chromatosome, and if chromatosome is further reduced, it is called the nucleosome core. Nuclease is an enzyme which decomposes the phosphodiester bonds of the nucleic acids. Nuclease is of two types, endonuclease and exonuclease. Exonuclease is mainly responsible for breaking the nucleotides from the ends of nucleic acids. Endonuclease on the other hand actually does the dispersing of the phosphodiester bond. There is another endonuclease which is known as the micrococal nuclease and is mainly responsible for breaking the nucleic acids into segments which are multiples of 200 base pairs. Each one of these 200 bp segment is attached with an octamer core of histone proteins and one molecule of H1 protein (where H1 is one of the classes of histone proteins). The proteins are used to prevent further action from the micrococal nuclease. But a strongly resistant structure is produced after H1 protein is lost and the length of DNA segment is 146 bp. This DNA segment is attached to the histone octamer core. And this particular structure is known as the nucleosome core. In a nucleosome, the DNA wraps 1.65 turns around a histone octamer consisting of two molecules each of histone proteins H2A, H2B, H3, and H4. A chromatosome contains two full turns of DNA (166 base pairs) locked in place by one molecule of H1. H1 histone proteins binds two nucleosomes together.

The diagram of the chromatosome is shown below:

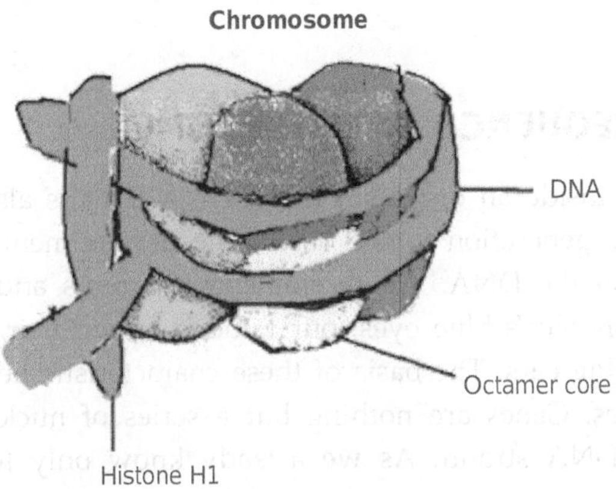

Fig. 6.11: The Chromatosome

Unit 7

NUCLEOTIDE SEQUENCE DATA

7.1 NUCLEOTIDE SEQUENCE DATA: GENOME

The genetic material inside an organism is the DNA. This is also the hereditary material which is passed from one generation to the other. The development and multiplication of an organism is dependent on the DNA. We resemble our parents and grandparents in having similarities such as, our mother's blue eyes, our father's brown hair, our grandmother's short height, our grandfather's big ears. The basis of these characteristic lies at the genetic level and this is what we call genes. Genes are nothing but a series of nucleotides arranged in some particular order on the DNA strand. As we already know only four nucleotide bases are present in the DNA (adenine, guanine, cytosine and thiamine). The genes make up for different sets of arrangements of these four bases. They may differ in length according to their complexity. The longest human gene is 2.4 million base pair long. An organisms entire stretch of DNA is what is called the genome, which contains all the genes and other non coding regions. In a human cell the genome is around 3 billion base pairs or nucleotides.

7.2 GENOME SEQUENCING

What we would like to focus on now is how a gene transforms into a phenotype or what function does it perform in the organism's body. Earlier approach in the field of molecular biology has been to look at a single gene at a time and analyze it for further study. But now with technological advances, many genes may be studied at the same time. Genome sequencing helps to find out the function of a genome, how genes are expressed and how do they transform into observable characteristics and also the similarities and the dissimilarities that exist among various organisms regarding their genomes. Scientist still rely on the traditional method of taking a single gene or protein at a time for experimentation and to completely comprehend the role of the gene and its biological significance. But as the databases promise to be expanded and store vast amount of information, the experimental strategies are also gearing up to take advantage of that and perform parallel gene studies. DNA sequencing is the process in which a molecule of DNA is taken and the order of nucleotides found out. A small amount of DNA may serve as big enough genetic material to sequence an entire genome. Usually the saliva cells or bone marrow cells or hair follicle cells are taken.

Genome sequencing is the process of finding out the exact sequence of the 3 billion nucleotides (abbreviated A, T, C, and G) that make up the genome of a human cell. The genome is nothing but the DNA of all the 24 different chromosomes taken together of a human cell. The genome sequencing is done by breaking down the genome into fragments of around 500 base pairs and then the sequenced fragments are reorganized in a consecutive sequence. The two most well known approaches for genome sequencing are: (1) Shotgun Approach and (2) Clone Contig approach. In general these two approached are mixed together to perform sequencing of larger genomes.

7.3 THE SHOTGUN APPROACH

For DNA sequencing, the shotgun approach appears to be the best method. In shotgun sequencing, the entire genome or a specified part of the genome is broken into arbitrary fragments. These fragments are then put into plasmids to make multiple copies of the fragments. A plasmid is a biological vector that can take any arbitrary portion of DNA and recreate copies of that instantly to provide enough matter for sequencing. When the DNA in the genome is of considerable size, the clones are made in such a way that they pass over the fragments of the genome pairs many times. The ends of the DNA clones are then numbered or in some cases, both ends are numbered. It so happens that a shotgun experiment requires an entire genome to be passed over by the clones for around 6-10 times. The last phase of the sequencing process requires that the numbered fragments be put back together into a sequence, which is commonly known as sequence assembly. While doing sequence assembly, it is observed that putting the small sequences together back again does not produce one full sequence. On the other hand it produces many clearly defined or distinct sequences that don't run over each other. These clearly defined sequences are known as *contigs*. In the assembling process, the beginning and stopping of contigs happen because certain regions are met where we do not get enough matter to further carry on assembling the fragments. These regions are known as gaps. The last step of the Shotgun process is to fill these gaps.

7.4 THE CLONE CONTIG APPROACH

The clone contig approach incorporates the shotgun approach for assembling but for cutting the genomic DNA into smaller fragments, it applies a different strategy. For cutting down, it identifies restriction patterns. The entire genome is not broken down into arbitrary fragments but into restriction fragments. The cloning of these restriction fragments is performed using chromosomal vectors which are artificial in nature. Many copies of the fragments are made and the fragments are numbered and then finally reassembled using shotgun approach.

The breaking of the DNA takes place with the help of some enzymes known as restriction enzymes. The restriction enzymes operate as follows. They cut the DNA only when they encounter a specific sequence of 6-10 base pairs. They cut at those 6-10 base pairs and in this manner the genomic DNA is not cut very frequently. The genome sequence of any organism is not arbitrary but particular to that organism, so when the genome is cut via his method, the consequence is that particular sequence patterns emerge.

7.5 SEQUENCE ASSEMBLY

The shotgun approach produces millions of fragments which are 400-500 base pairs in length. The fragments are arbitrary in nature and overlap each other fully or partly. The overlapping nature helps to identify fragments which lie next to each other, because the sequence of some nucleotides at the end of one fragment resemble the sequence of nucleotides at the beginning of another fragment.

The Process of Sequencing using Shotgun Method

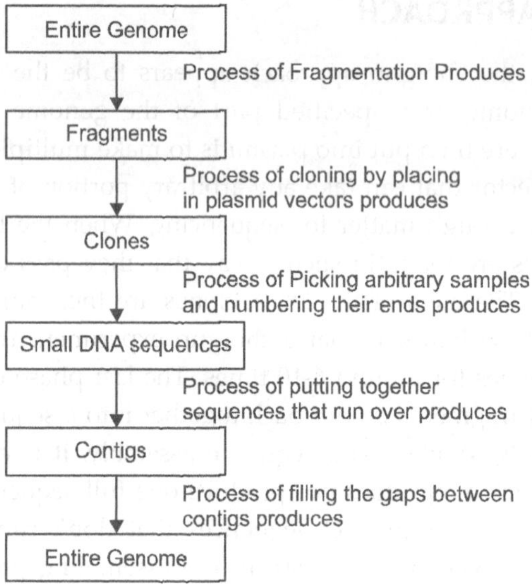

Fig. 7.1: The Shotgun Approach

7.6 RESTRICTION ENZYMES

Restriction Enzymes are made in the bacteria and borrowed from the bacteria for use in molecular biology. These restriction enzymes can be separated from the bacterial cells and taken.

The role of a restriction enzyme is to cut the double stranded DNA at a specific site known as the restriction site. The restriction site is nothing but a particular sequence of nucleotides arranged in some specific order. Whenever the enzyme meets these restriction sites, it cuts the DNA molecule at that point.

The very first enzyme identified as restriction enzyme was found in E Coli, known as *EcoRI*. The restriction enzyme is also known as restriction nuclease. The discovery of restriction enzymes has helped in the field of medicine. For example, Insulin was produced in vast quantities so that diabetic patients could be treated. Insulin is a result of recombinant DNA (rDNA). A recombinant DNA is nothing but the joining of two DNA strands when they have been cut through a restriction enzyme. Around 3000 restriction enzymes have been analyzed so far and more than 600 are used for commercial purposes.

The EcoRI cuts the DNA whenever it meets the sequence "GAATTC" in that order. The restriction enzyme pierces the DNA twice, once through the sugar-phosphate backbone of one strand and another through the sugar phosphate backbone of another strand. One strand of DNA is cut at one point and the second strand is cut at another complementary point. The isolated parts of the DNAs are known as "sticky ends". These sticky ends can be joined to another sticky end which has the complementary base pairs which can attach to the base pairs of the first sticky ends. They are further fixed to each other for ever with the help of another enzyme called DNA ligase. The DNA ligase forms covalent bonds or joins the phosphodiester bond along the backbone. These combined pieces from different DNAs are what is known as the recombinant DNA.

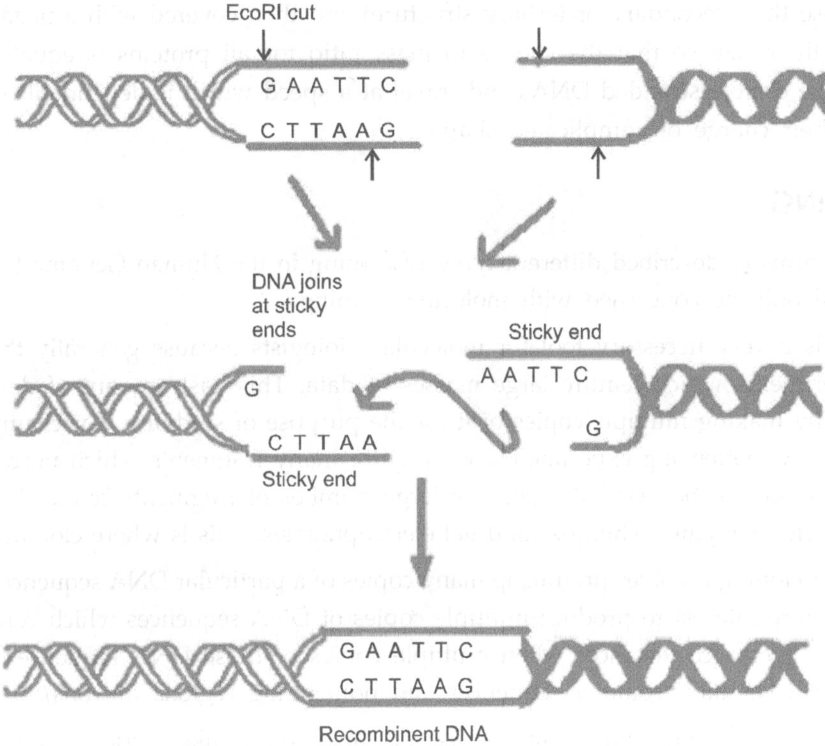

Fig. 7.2: The Function of Restriction Enzymes

7.7 GEL ELECTROPHORESIS

Gel electrophoresis is a process in which DNA, RNA and protein molecules are placed in a gel matrix and then isolated from each other when electric current is passed through the gel matrix which is porous in nature. The gel's make up and permeability is decided according to the weight of the DNA, RNA or protein molecules. For smaller molecules the gel is made up of acrylamide for larger molecules the gel is made up of agarose.

The way in which gel electrophoresis works is that molecules are put in different wells in the gel. On passing the electric current the molecules travel at different speeds depending on their sizes. The heavier ones move slowly and the lighter ones move faster and therefore at the end of the experiment they are all placed at different positions in the gel.

To see the molecules generally dyes are used. Another alternative is to use ultraviolet lightning conditions. Another method is to attach the molecules with radioactive substances.

The nucleic acids have a negative charge due to the sugar-phosphate backbone, therefore the molecules move from a negative pole to a positive pole. Double stranded DNAs move according to their size. Single stranded DNA or RNA fold to form a tertiary structure and move in a complex way according to the shape they take.

The proteins differ with respect to their charge and complicated shape. They are therefore denatured (lose their secondary or tertiary structure) and then covered with a negative charge according to their size so that the charge to mass ratio for all proteins is equal. They then behave just like double stranded DNAs and travel at a speed which is dependent on their size and not on their charge or complicated shape.

7.8 CLONING

We have already described different types of cloning in the Human Genome Project topic. Here we shall only be concerned with molecular cloning.

Cloning is a very necessary tool for molecular biologists because generally they need to perform experiments which require large masses of data. This vast amount of data can only be generated by making multiple copies of it for the purpose of studying. For example, for the purpose of DNA sequencing experiments, we require many fragments which need to overlap so that their order can be decided later. The large number of fragments cannot be generated only by restriction enzyme techniques and gel electrophoresis. This is where cloning is helpful.

Molecular cloning involves producing many copies of a particular DNA sequence. Generally it may be of more interest to produce multiple copies of DNA sequences which contain genes, but in practice there are instances when multiple copies of those DNA sequences have been produced which contain promoter sequences and non-coding regions (introns).

The process of cloning mainly makes use of chromosome like carriers called vectors in which the specific DNA fragments are inserted and their replication takes place in isolation to produce identical copies which are known as molecular clones and they can later be purified for testing or stored in libraries for future use.

The vectors must have the following features:

1. They should be able to replicate inside the living cell.
2. Their presence can easily be detected.
3. Their size and shape should be different so that they can easily be separated from the host cell's DNA.

A genomic library is the one which contains every fragment of the DNA in the entire Genome covered at least once. A genetic library is a collection of genes which have been formed after cloning. A genomic equivalent is defined as the genome size divided by the average fragment length which gives a rough idea of the number of clones that need to be present in the genomic library.

For example if E.Coli's genome length is 4,600,000 nucleotide long and the average DNA fragment length is 4,096 base pairs then the number of clones required would be 1000 and that would be called the genomic equivalent.

The cloning process is not an easy task normally it takes hundreds and thousands of cells to clone a single fragment. The cloning process also does not ensure that the entire genome would be cloned at least once. Many fragments may be repeated multiple times and there would be fragments which have not been cloned at least once. We can only increase the chance that every DNA fragment will be cloned at least once by increasing the no of clones.

In a number of cases the cDNA libraries are also found as an interesting alternative to the cloning process. Normally we are interested in those genes which code for proteins. Before the process of translation the mRNAs are separated from all the other polynucleotides within a cell with the help of an enzyme called reverse transcriptase, which converts them back into complementary DNA (cDNA) and these cDNAs are collected and stored together into a cDNA library.

7.9 POLYMERASE CHAIN REACTION

When more detailed analysis is required and little information is available which needs to be transformed in larger quantities then the method of Polymerase Chain Reaction (PCR) becomes useful and powerful when compared to cloning.

In this one double stranded DNA is taken. It is heated upto 95 °C to denature it, and then the temperature is cooled down to 55 °C and the primers are allowed to anneal to it, which means the temperature is first raised and then slowly allowed to cool. Then polymerization takes place in the sense that the primer extends in which DNA polymerase is used. This is done at a temperature of 72 °C.

To the denatured single stranded DNA's attach the primers and extend to make the target DNA. At the beginning of the 2nd cycle, the two copies are used to make 4 copies and in the 3rd round they become eight. So in the chain reaction one target molecule becomes 2^n molecules after n cycles. Normally 20–30 cycles are used to achieve sufficient large quantities of DNA.

This method is particularly helpful when smaller quantities of material are available. These can be taken from forensic specimens or from ancient biological samples or from samples of biological laboratories.

Normally PCR primes are 20 or more nucleotides in length. The PCR primers are designed to anneal on opposite strands of the target sequence so that they will extend towards each other by addition of nucleotides to their 3′ ends. Normally Taq Polymerase of DNA polymerase sustains a temperature of 95 °C.

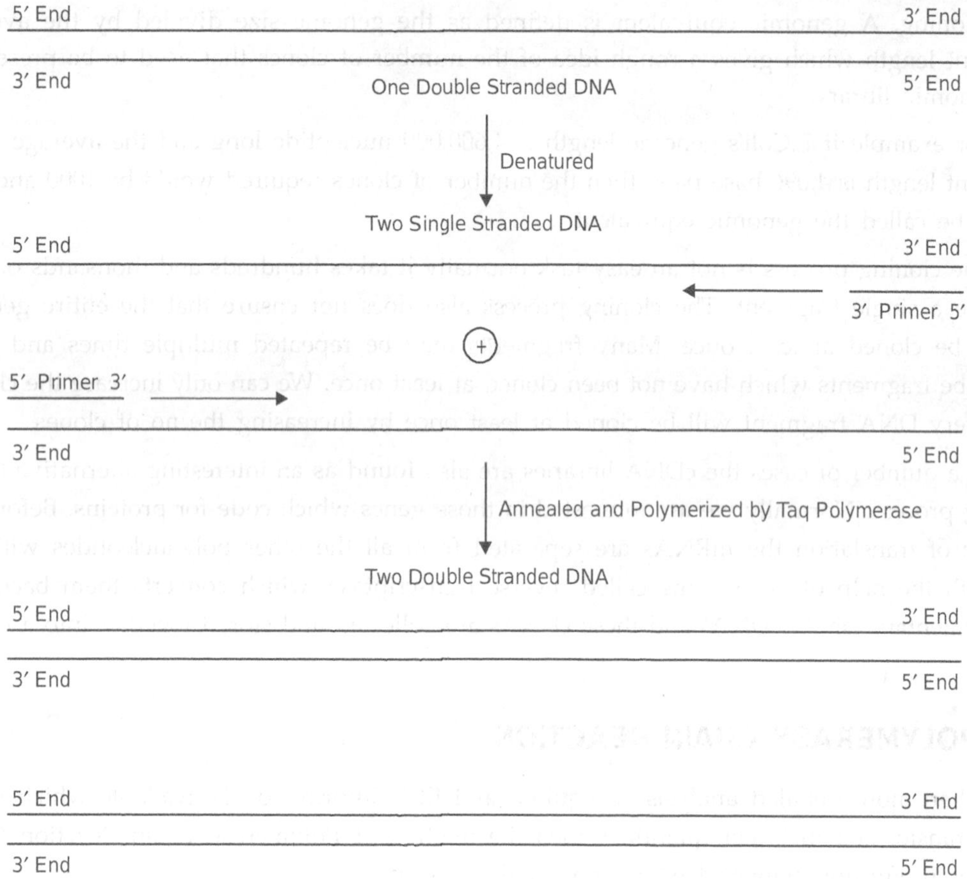

Fig. 7.3: The Polymerase Chain Reaction

7.10 DNA SEQUENCING

The ultimate method in which the order of nucleotides can be found is known as the DNA sequencing method. It uses any method or technology that is used to determine the order of the four bases – Adenine, Guanine, Cytosine and Thymine in a strand of DNA.

There are 2 main methods with which we do DNA Sequencing and they are:

1. Maxam and Gilbert Method
2. The Sanger's Enzymic method

Each of the two main methods of sequencing long DNA molecules involves the production of a set of different sized molecules with one common end which are then separated by polyacrylamide gel electrophoresis to allow reading of the sequence.

In the Maxam and Gilbert Method, it allows the DNA fragment to be grown by adding first a chemical called piperidine. Incubation time is given so that the ladder grows. Then with the help of chemical degradation the fragments are broken. Four chemicals are added to the mixture. Dimethyl sulfate attacks the G bases. Formic acid attacks the purines A and G.

Hydrazine is used to attack pyrimidines (C + T) and high salt stops the T base. Thus four lanes are made in the Sequencing Gel, (G, A + G, C + T, C) and the sequences determined.

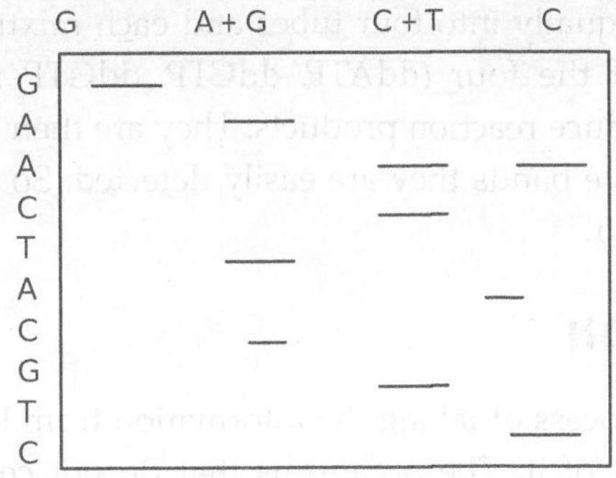

Fig. 7.4: The Maxam and Gilbert Method

7.11 SANGER CHAIN TERMINATION METHOD

The Sanger approach is also known as the chain termination Method because in this modified nucleotides are added which stop the DNA polymerase to grow and add additional bases at their ends. These are known as the chain terminating nucleotidesand the reason that they don't let grow the chain is that their Hydroxyl group (OH) is missing on the 3′ end which mainly helps in adding the nucleotides to its end.

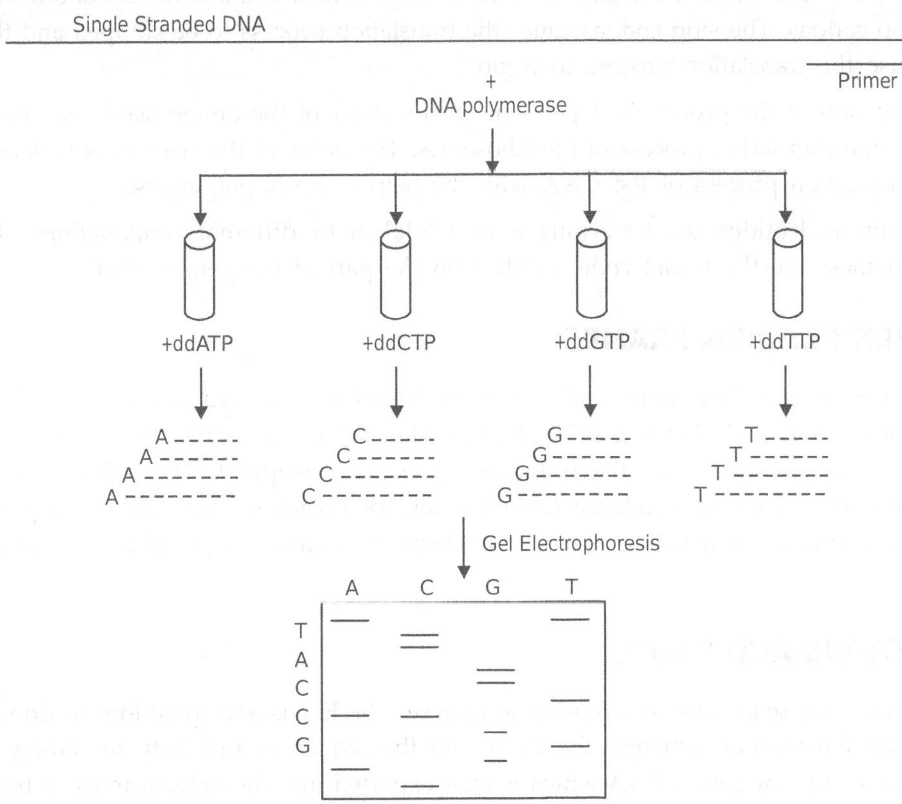

Fig. 7.5: The Sanger Chain Method

First a single stranded DNA with unknown sequence is taken as a template in that is added a radioactively labeled primer and DNA polymerase is also mixed to let the chain grow. This extended mixture is divided equally into four tubes and each mixture is given a different chain terminating nucleotide out of the four (ddATP, ddCTP, ddGTP and ddTTP). Then they are allowed to synthesize and produce reaction products. They are then passed to Gel electrophoresis and with the help of radioactive bands they are easily detected. So all four of them stop at their respective bases A, C, G and T.

7.12 GENE EXPRESSION

Gene Expression is the process of taking the information from DNA, converting it into RNA and then making proteins out of it. The organisms that do not code for the right proteins do not survive compared to the ones that express their genes properly. In order to do proper gene expression, the organisms should be able to make out which genes should be expressed and at what time and also the starting location of those genes.

7.12.1 The Genetic Code

The genetic code can be defined as the order of the nucleotides which are arranged on the DNA molecule. Since early history of mankind the same genetic code is still used by all living things till today. The genetic code is defined by codons which is a set of 3 nucleotides. There are generally the same set of 3 nucleotides used as start codons and another set of 3 nucleotides used as stop codons. The stop codons cause the translation process to be stopped and the start codons cause the translation process to begin.

The function of the protein is dependent on the order of the amino acids that are strung together by the translation process of the ribosomes. The order of the ribosomes is determined by the transcription process of the DNA with the help of RNA polymerase.

The four nucleotides can be arranged in a total of 64 different combinations of three. $4^3 = 64$ and these are the triplet code which form the part of the genetic code.

7.13 OPEN READING FRAMES

There is only one start codon and that is the triplet AUG stored in this order. There are 3 stop codons and they are UAA, UAG and UGA. The start codon starts the translation process and when a stop codon occurs the translation process is stopped. The uninterrupted part between the start and the stop codon is known as the open reading frame and the open reading frames also comprise of a st4ing of codons which are mainly responsible for making the proteins.

7.14 INTRONS AND EXONS

Introns are the sequences which occur in between the Exons and these Introns do not help in coding for the proteins whereas the Exons are the sequences that help in coding for the proteins. To create a mature mRNA which is later exported into the cytoplasm to be translated into proteins is first present in the nucleus in the form of pre-mRNA. The pre mRNA consists of sequences of Introns and Exons joined in any order. The processing of pre-mRNA takes in the form of splicing in which the Intron sequences are removed and the Exon sequences are joined to form the mature mRNA. Intron sequences are degraded once they are removed.

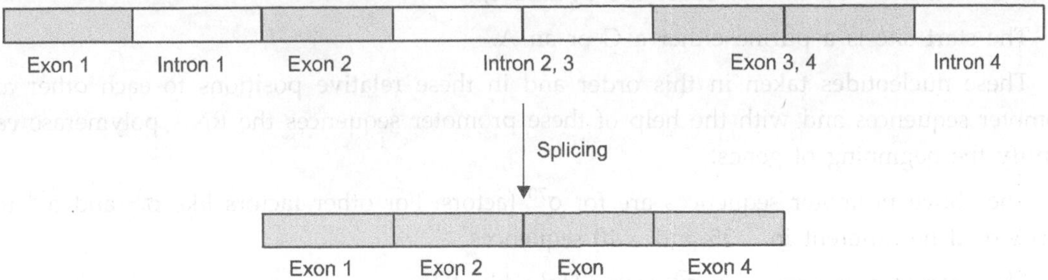

Fig. 7.5: The Process of Splicing

7.15 CONTIGS & EXPRESSED SEQUENCE TAGS OR ESTS

In order to define Expressed sequence tags it is important to define a term called ***Contigs***.

A single chromosome generally requires many sequencing reactions in order to give it a complete coverage. Contigs are defined as continuous runs of nucleotides longer than what can be obtained from any single sequencing reaction. From a number of multiple sequencing reactions the areas that show significant overlap by taking the sequencing reactions above the minimum number are defined as contigs. On the chromosome exist sites which are also known as sequence tagged sites and they are the ones which show most significant overlaps. These are called as ***Expressed Sequence Tags*** or ***ESTs*** and a no of ESTs may be combined to form Contigs.

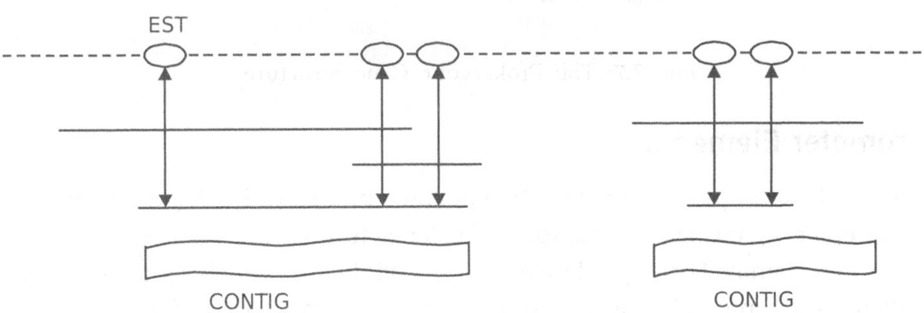

Fig. 7.6: Contigs and ESTs

7.16 PROMOTER ELEMENTS

To begin the process of transcription, when RNA polymerase binds to specific conserved sequences, these are called promoters. Promoters are basically where the RNA polymerase looks to bind itself to start transcription.

In the prokaryotes, RNA polymerase scans the DNA looking for approximately 13 nucleotides in that order. 1 nucleotide marks the beginning ans is known as transcriptional start site, 6 bp regions which begin 10 bp upstream from the start site (and this is known as the – 10 sequence and the six bp are TATAAT in this particular order). Then it requires another 6 bp regions which are 35 bp upstream from the start site (and this is known as the – 35 sequence and the 6 bp in that order are TTGACA).

The start site is a purine either a G or an A.

These nucleotides taken in this order and in these relative positions to each other are promoter sequences and with the help of these promoter sequences the RNA polymerase can identify the beginning of genes.

The above promoter sequences are for σ^{70} factors. For other factors like σ^{32} and σ^{54} the 6 bp would be different in – 35 and – 10 sequences.

The promoter sequence for σ^{70} factor looks like this:

TTGACA	----------	TATAAT	--------	G or A
– 35 sequence	16 – 18 bp	– 10 sequence	5 – 8 bp	Start nucleotide

7.17 PROKARYOTIC GENE STRUCTURE

A Prokaryotic Gene Structure looks like this:

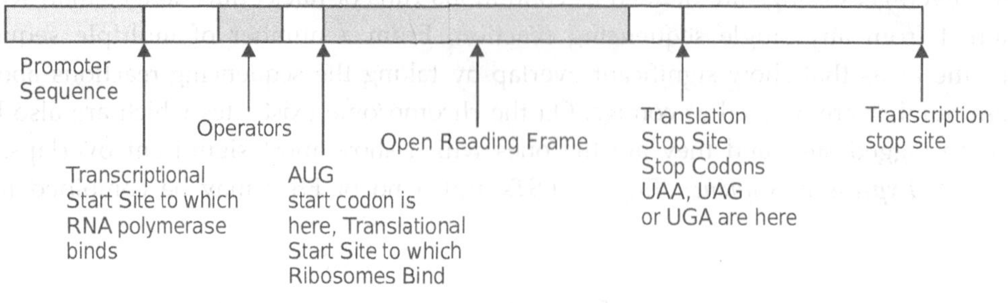

Fig. 7.7: The Prokaryotic Gene Structure

7.17.1 Promoter Elements

The responsibility of promoter recognition lies on a protein called σ (Sigma). It has different variants in different organisms, for example σ^{70} belongs to general gene family and σ^{32} belongs to Heat Shock gene family. For example when in E. Coli due to heat shock the genes need to be expressed then the σ^{32} promoter sequences are located and their particular – 35 sequences and – 10 sequences are found which are together known as consensus sequences and the process of transcription begins.

σ factor	Gene Family	–35 Sequence	– 10 sequence
σ^{70}	General	TTGACA	TATAAT
σ^{32}	Heat Shock	TCTCNCCCTTGAA	CCCCATNTA

An operon sequence is known to have more than one promoter and arranged in a sequence. The promoters can act as positive or negative regulators. A negative regulator restricts transcription and a positive regulator promotes transcription. In an operon, a negative regulator may be on one side which will prevent further transcription from that end. An example of a negative regulator is lactose repressor protein(pLacI) then in between may be the normal promoter sequence and at the other end may bind a positive regulator such as (CRP) Cyclic AMP Receptor Protein which promotes further transcription at the end, so three promoter sequences are found together in one sequence called the operon.

7.18 EUKARYOTIC GENOME

Eukaryotic genomes are much more complex and large in size as compared to the prokaryotes. A eukaryotic cell has the ability to maintain variety of chemical environments within it as compared to a prokaryotic cell which maintains a single environment. The DNA of each cell in a eukaryotic organism may be the same but different cells express different genes. The genome sizes of eukaryotes are extremely large therefore the "Junk DNA" present also tends to be large. Prokaryotes contain a single circular chromosome in their cell whereas Eukaryotes have 2 copies of the same chromosome for example humans have 23 chromosomes out of which there are 2 copies of same chromosome 1-22 and one sex chromosome is X in females and Y in males.

A human genome has approximately 3,200 million basepairs and around 45,000 genes, a small eukaryote (yeast) has 13.5 million bp long genome and around 6,241 genes. Fruit flies or Drosophila Melangaster has around180 million bp of genome size and 13,601 genes which can be expressed.

So the size of the genomes of the eukaryotes varies upto greater extent from small yeast like structures to some plants and to human and blue whales, to generalize the genome size is not possible.

7.19 EUKARYOTIC GENE STRUCTURE

In Eukaryotes, the no of introns are quite high and their promoters do have some conserved features but are located far apart from the starting codons. In eukaryotes the problem of locating the genes is amplified and remains a great challenge for bioinformaticians.

In Eukaryotes, the same genetic information exists in all cells, a kidney cell also has the same information as the heart cell then how are they different. They are different because they have expressed their genes differently.

In prokaryotes we use to use one RNA polymerase to initiate transcription. Due to the large size of genomes and a variety of genes in eukaryotes 3 types of RNA polymerases are required :

1. RNA Polymerase I
2. RNA Polymerase II
3. RNA Polymerase III

Each one of these RNA polymerases require a different set of promoters and they all express different kinds of genes.

RNA Polymerase I and RNA Polymerase III express those types of genes which are functionally important for the organism and are required at all times at about the same levels.

Out of these RNA Polymerase II is the one which codes for proteins. Now it is the sole responsibility of RNA Polymerase II to recognize the promoter sequences to be able to find out which gene needs to be expressed at what time in which type of cell.

Every Eukaryotic gene has its own promoter.

The RNA Polymerase II begins its transcription where it finds its first set of promoters called the basal promoters, where the RNA Polymerase II initiation complex is assembled and transcription begins.

Apart from the basal promoter other type of promoters known as upstream promoter elements are found to which proteins other than RNA Polymerase II bind. To uniquely identify a gene a minimum of 5 upstream promoter elements are required so that it is expressed properly.

In eukaryotes the basal promoters are not enough to recognize the starting of genes by RNA polymerase II. Instead basal transcription factors are required which includes a TATA binding protein (TBP) and at least 12 TBP associated factors (TAF), these together seek the nucleotide sequence of promoters in specific order and then help RNA Polymerase II to bind.

The TATA box has a consensus sequence of 5 – TATAWAW – 3 where W means either A or T is present at – 25 instead of – 10 relative to the transcriptional start site (+ 1). In eukaryotes the +1 position is associated with an Initiator (Inr) Sequence (with a consensus of 5′ – YYCARR – 3′ where Y means C or T and R means G or A.

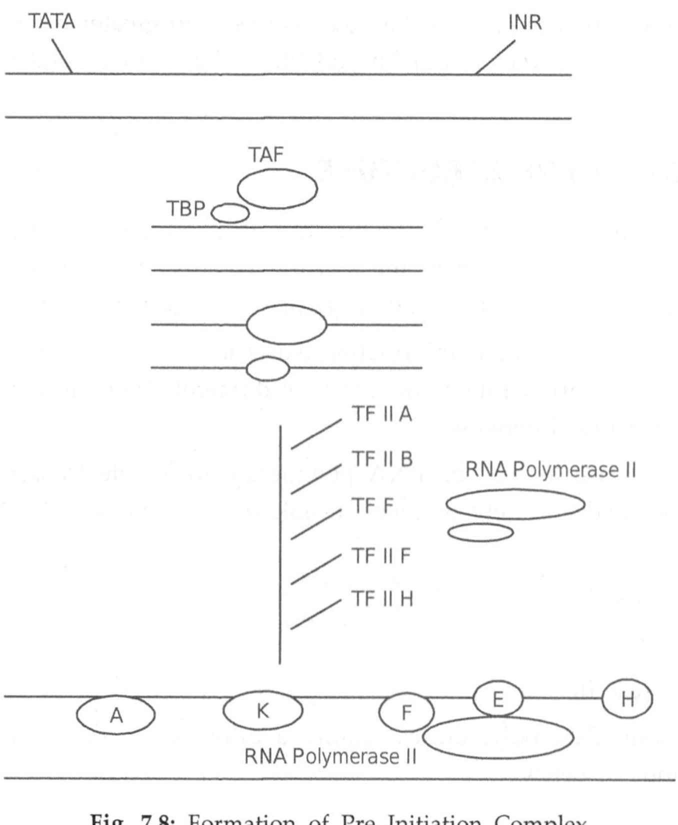

Fig. 7.8: Formation of Pre Initiation Complex

7.20 REGULATORY PROTEIN BINDING SITES

To regulate the protein binding sites there are positive regulators which work on many genes and do not respond to external stimuli which are also known as constitutive. Other type of regulators are called regulatory which work on small no of genes and do respond to external stimuli. These constitutive and regulatory activating proteins are called the transcription factors.

Most transcription factors identify the specific DNA sequences and bind to them in the range of -80 bp and are relatively close to the transcriptional start site. Other transcription factors which are around – 500 to + 500 bp away from the transcriptional start sites are known as enhancers. Enhancer sequences are typically 100-200 bp long and contain multiple sequence elements which contribute to the total activity of the enhancers. Some enhancers work tens of thousands of bp upstream or downstream of the transcriptional start sites and they tend to bend the DNA into certain shapes so that other transcription factors can be brought into contact with each other to form structures known as enhanceosomes. Other transcription factors are responsible for expression of genes under certain conditions at certain times in specific tissues.

7.21 OPEN READING FRAMES

While the RNA is transported out of the nucleus to begin the process of translation. This delay helps the eukaryotes to change their RNA polymerase II transcripts in which the process of capping, splicing and polyadenylation takes place.

In capping the chemical alternation of the heterogeneous hnRNA takes place. In splicing the endless no of stop codons which interrupt the ORF (Open Reading Frames) are cut so that gene expression can be done properly and here the tissue specific gene expression can be differentiated and finally the process of polyadenylation which replaces the 3′ end of an hnRNA with approximately 250 A's.

7.22 INTRONS AND EXONS

The hnRNA is converted into mRNA from one prominent intron found in eukaryotes is the GU –AG rule. It says that the first two nucleotides at the 5′ end of the DNA sequence of all of these introns are invariably 5′ – GU – 3′ and the last two at the 3′ end of the intron are always 5′-AG -3′.

Introns need to be at least 60 bp long so that they can be found out by splicing signals and human genomes contain introns with the upper limit of many tens and thousands of base pair long sequences.

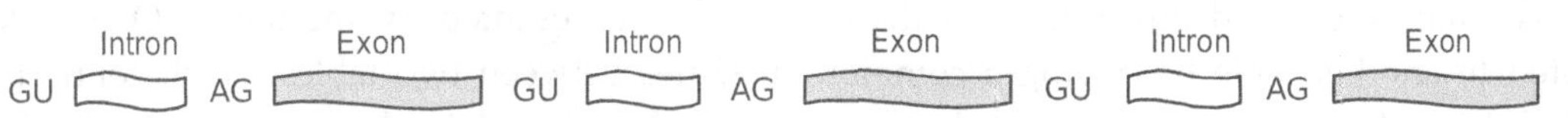

Fig. 7.9: Introns and Exons

7.23 ALTERNATIVE SPLICING

When splicing occurs then the majority of eukaryotic genes are spliced into a single type of mRNA.

Alternative splicing occurs when more than one type of mRNA is produced from the same gene. This happens because when hnRNA is converted into mRNA then not all the same exons will be expressed at all times under all circumstances.

The splicing signals are given by very small nuclear RNAs called the (snRNAs) as well as several proteins each of which differs from one cell type to another. Different splicing signals are recognized in different tissues and therefore same set of genes do not give rise to same set of proteins and hence the change occurs.

7.24 DATABASES

Bioinformatics data is stored in large data bases like GenBank in terms of nucleotide sequences.

Still a DBMS hides the underlying technology and the hardware from the user. In order to access and manipulate the data, the data must be organized in proper way so that it can retrieve the information in an efficient way.

There are many models on which databases are based such as flat, relational, hierarchical, network and object oriented. Here we will be discussing the flat, relational and object oriented databases in relation to Bioinformatics data, their advantages and disadvantages.

7.24.1 Flat File Databases

Normally the type of information that would be required to store in the databases of Bioinformatics would be protein sequences, protein structures, mRNA, genes, nucleotide sequences etc. So we require a type of a database model which best captures all these entities and shows correct relationships between them without messing up with unnecessary duplicate or redundant data and unable to correct the entities properly.

In a flat file system there is simply a table without any embedded structure information to associate the relationships between records. The limitation of a flat file DB is that it can only work with a single file or with a single table at any one particular time. Also another limitation is that no embedded structures can be associated with it.

The advantage is that it is able to store large sequences of data easily without showing any relationship between them, so it becomes useful in storing nucleotide sequences efficiently.

7.24.2 The Relational Model

The Relational Model is based on rows and columns of a table, the rows are called tuples and columns are called attributes and the entities are represented by the tables. One table is related to another table by having a common attributes between two tables which connect the tables. Records with a common attribute can be combined or joined with any two files.

The limitation is that it takes huge amount of time to retrieve data as it takes a lot of time to perform the join between the tables.

The advantage of relational model is that it is reliable, the concepts can be widely understood, the entities can easily be related to one another, the data can be populated in rows and columns of the table and matching fields between two tables are joined to make larger tables and to associate one entity with another.

The limitations are it mostly requires all rows and columns to be populated, also all rows and columns need values oriented data. Then if we want to relate DNA sequence with amino acid sequence then we require an inter mediatory protein sequence which will connect with the DNA sequence with one table and amino acid sequence with another table and only then the DNA and amino acid sequences will be related.

Then Relational Model always works with a table of rows and columns which is not a very natural method for data of bioinformatics type because bioinformatics data is more abstract connected in abstract ways not always fixed into rows and columns.

7.24.3 The Object Oriented Model

So the answer to the above problems is to use the object oriented model in which a number of combinations can be presented, first hierarchical relationships can be shown easily then the type of data need not always be values, they can be structures, shapes, images, data values, sequences etc.

Here one object may contain another object, these objects may be of any type, the relational and the hierarchical model can both be fitted into this. Arbitrary data and complex relationships can be forwarded easily. No complex joins between tables are required, data can naturally be connected to the type it needs to connect.

This model is considered best for handling Bioinformatics data.

For example there is an object gene which has introns and exons, then those exons have protein sequences and amino acid sequences, this can be shown as

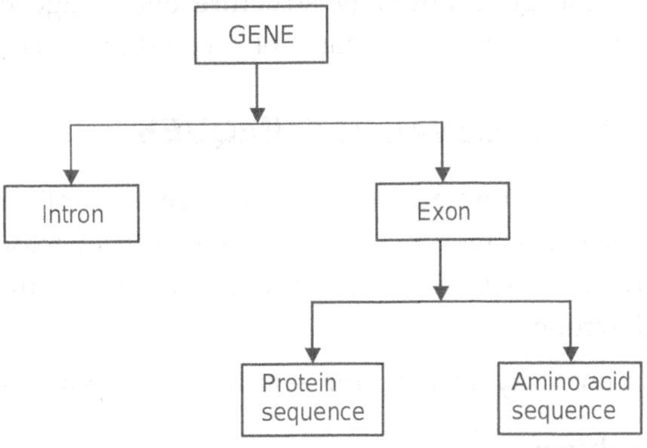

Fig. 7.10: The Object Oriented Model

All these are objects within objects and can hold any type of data.

The limitation of object oriented approach is that users are more comfortable using the relational database model and also relational database has a large global market. Object oriented database users find difficult to use and comprehend, so most commercial packages use underlying database as relational with extended features of object oriented data type fitted into it.

7.25 CONTROLLED VOCABULARIES

The vast integration of information from various sources today have led to a challenge which is faced by many Bioinformaticians today and that is to refer to the same entity by same terms in all occasions of use.

Thus has led to the development of a public database known as Gene Ontology (GO) which is designed to create a list of biologically relevant and structured terms that can be shared among all sorts of bioinformatics resources.

The controlled vocabulary terms describe gene product characteristics such as biological methods, molecular roles and components of the cells.

GO can be browsed and queried in many ways. Many genome databases use GO to add important functional notes of explanations to their characteristic genes.

When a gene is identified, three important questions need to be addressed:

1. Where it is located in the cell
2. What function does it have on the molecular level.
3. What contribution do these functions make in the biological processes.

The function of the GO database is to capture the relations between biological processes and the molecular functions which underlie them.

The terms become unified indirectly where single gene products are simultaneously marked with explanatory notes via terms from different heterogeneous sources.

An understanding of the gene ontology structure and usage can help researchers to perform better queries and to characterize many of biomedical research data types.

7.26 GENERAL DATA RETRIEVAL TECHNIQUES

Data retrieved from the Bioinformatic databases can give us information which is relevant and also information which is irrelevant. Searches may retrieve too much noise with the relevant data or they may also retrieve the wrong data, because of the search criterion which may have been inputted wrongly.

Boolean searches are one way to limit the selectivity of data.

7.26.1 Boolean Searching

The Boolean operators used are AND, OR and NOT.

The OR operator provides the least amount of selectivity, where as the AND operator provides the greatest amount of selectivity. The NOT operator generally provides intermediate amount of selectivity.

A non-selective search which mainly uses just general terms tend to retrieve a large no of irrelevant data. And if we make the search selective, it will retrieve relevant data but may miss out on some of the desired data.

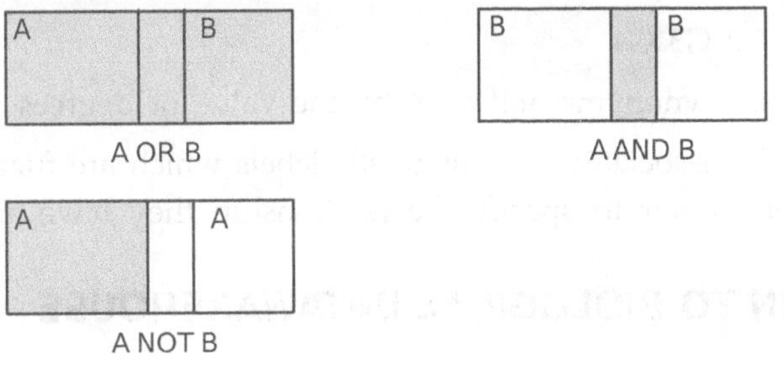

Fig. 7.11: Boolean Operators

7.26.2 Indices Based Searches

Some RNAs require to match sequence structure patterns in sequence databases.

There is a method which uses affix arrays which is an index data structure preprocessed from the target database.

Affix arrays are able to do pattern search in 2 directions which is mainly required for handling the structural constraints of the pattern.

Structural patterns are stem like loops which can be matched inside out. In this the loop region is matched first and then the matching on the boundaries is done. Then the matched substring is extended to the left and to the right, this is known as the bidirectional search.

In the sequence structure search, there are few sequences to be matched but more of structural and functional properties are also taken into account which need to be matched.

The indexes are made in such a way that they support either unidirectional search or bidirectional search.

The unidirectional search is the normal left to right traversal of the data structure.

In bidirectional, the character comparisons are started at any position of the pattern extending the already matched substring of the pattern to the left or to the right.

7.26.3 Fuzzy Searching

Fuzzy searching mainly targets to develop precise language that is not vague and could provide an optimal methodology in which terms from various species can be categorized.

Fuzzy linguistic terms are used to develop a new language to specify rules for bioinformatics classification.

The fuzzy matching is used to provide a tool in which queries can be evaluated for their imprecision of data. This is done by labeling the linguistic terms in queries and in classification.

Fuzzy sets can map the numerical values with linguistic terms by using membership functions and fuzzy pattern matching can be used to draw a comparison between the terms.

For example suppose a Swiss Prot entry has

An Accession Number: P13654

Entry Name: IL313_Fish

Protein Name: Leukin

Has 4 Genes : G1 G2 G3 G4

The above genes are synonyms and vary by the value of degrees present in them.

These values may be associated with linguistic labels which are fuzzy based such as very similar, little similar or similar to specify the relationship they have amongst themselves.

7.27 APPLICATION TO BIOLOGICAL DATAWAREHOUSE

The Datawarehouse may be described in terms of databases, data repositories and data marts.

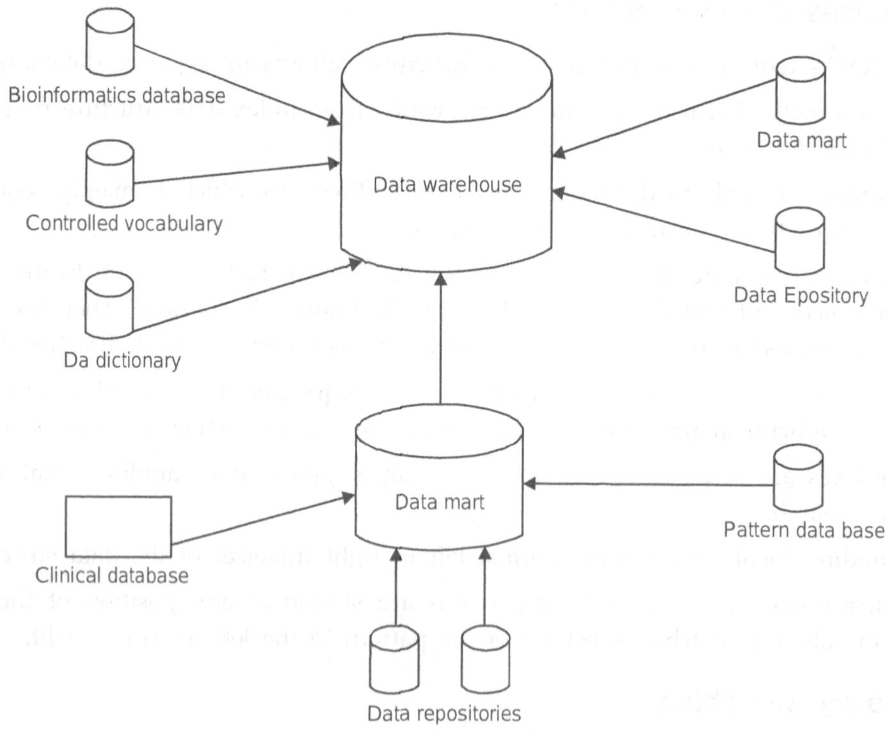

Fig. 7.12: Biological Data Warehouse

At the lowest level of a datawarehouse lies a data repository, which is a database that stores some information with minimum querying facility. A data repository is generally built from a single source of information such as clinical data or microarray.

The data repository is different from the database in the sense that data can be stored in duplicate repositories and while querying it need not disturb the DBMA, it can be offline.

After the data repository comes the data mart, which is focus oriented in the sense that it collects the data from a number of repositories and application databases to store information about some specific task. It is data extracted for a particular purpose.

The data warehouse at the top most level is formed from a number of sources, a number of data marts, controlled vocabularies, data dictionaries, pubic bioinformatics databases.

The data comes from heterogeneous sources therefore it is very important to have one common format or language which could be understood by all and that is where the controlled vocabularies come into picture, they are standardized vocabulary which maps identical concepts which are expressed in different words, phrases or units. A popular controlled vocabulary is the Medical Subject Heading (MeSH) maintained by the U.S. National Library of Medicine and used by PubMed Biomedical Literature Database.

The datawarehouse combines data from various sources so it is required to clean it, encode it and translate it into a standard format.

The data is then extracted, transformed, loaded, merged and put to a datawarehouse from where the Mining process starts. Now here in mining interesting useful patterns can be found which can give us an insight of the large amount of information stored in a datawarehouse.

Unit 8

Biological Data Types with Requirements

8.1 SEQUENCE COMPARISON

Sequences are compared by aligning them together and then finding the differences among them.

There may be single sequence alignments or multiple sequence alignments. Sequence comparison is mostly done of nucleic acids, proteins and RNAs. There are a number of algorithms which are used to do this and dynamic programming method is an advancement over the others. In comparing two sequences their distance and similarity is noted. The gaps are found out. Then pairwise alignments can be done or matrices can be formed to find out the scores. BLAST and FASTA are used for database searches, we will be discussing all the points related to sequence comparison in the following sections.

8.2 HOW THE SEQUENCES ARE ALIGNED

For any sequence comparison method, the first thing that we need to do is to find out how sequences are aligned together.

Sequence alignments in general give us an idea of how genes are related together, the functions of genetic sequences.

The sequence alignment generally is done for comparing DNA sequences or RNA sequences or Proteins in the hope of finding any similarities that will lead us to find interesting relationships in either their structural configurations, evolutionary relationships or the functions that they perform.

Sequence alignment is usually shown as rows in a matrix, where gaps may be inserted in between to show matches where pairwise nucleotides match.

For example

ACGTATCGTATTAGCCTATTAAGCCGCG

ACGT - - - - TATTAGCC- - - - - - GCCG - -

The areas where the base pairs match show that the regions are conserved that is from their ancestor they may have a structural or functional role and where there are gaps, they indicate that single or multiple mutations might have happened in the form of insertion or deletion of bases.

8.3 SIMPLE ALIGNMENTS

Simple alignments consist of very short or very similar sequences which can be aligned by hand and human eye is easily able to detect.

The problem arises when we have very large sequences which are not possible by hand and require computing power, for that various algorithms are developed. The alignments may be divided into 3 categories:

1. Local Alignments
2. Semi – Global Alignments
3. Global Alignments

In local alignment, the sequences are found within large sequences which are very different in nature and overall small enough to be handled easily.

Local alignments are often difficult to identify because regions of similarity may be difficult to find out but are preferable because they may show interesting patterns quickly.

The global alignments, the query sequences is matched from the beginning to the end giving the best possible match after tying out all alternatives.

A semi global alignment is like a global alignment but penalty free gaps are allowed at the beginning and end of the alignment.

In certain cases, when we align the sequences the gaps may be allowed or may not be allowed.

In simple alignments if gaps are not allowed then subsequences may be matched to original sequences and scores can be found out. The sequence with the largest score may be selected.

The scores may be found according to the following criteria:

1. If there is a match between the base pairs then add 1
2. If there is a mismatch between the base pairs then assign a value of 0.

And take the sum of all the matches found.

For example

```
GCTACGT          GCTACGT          GCTACGT
GCTAT              GCTAT            GCTAT
```

The first sequence match gives a value of 4. The second a value of 0 and third a value of 1.

8.4 GAPS

The above sequence was without gaps but gaps can be introduced to take care of insertions and deletions or mutations that would have occurred during evolution.

Suppose we introduce 2 gaps in the above sequences, then there are a number of ways in which the base pairs can be aligned.

A few examples are shown here:

GCTACGT	GCTACGT	GCTACGT	GCTACGT
GCTA - - T	GC - - TAT	GCT - - AT	GC - TA - T

If we have 1 for a match

0 for a mismatch

We can also introduce – 1 for a gap penalty

So the scores for the above 4 alignments become

1. 5 – 2 = 3 (5 matches, 2 gap penalty)
2. 3 – 2 = 1 (3 matches, 2 gap penalty)
3. 4 – 2 = 2 (4 matches, 2 gap penalty)
4. 3 – 2 = 1 (3 matches, 2 gap penalty)

8.5 SCORING MATRICES

If a base pair does not match and it is single, we can say single point mutation has occurred.

If on the other hand gaps are occurring due to insertions or deletions in the sequences then we say that an indel (insertion/deletion) event has taken place.

Also as discussed in earlier chapters there are amino acids which have some specific properties such as charged, polar, hydrophobic, etc.

If there is a mutation where a charged amino acid was replaced by a charged amino acid then it should be given more advantage than if a charged amino acid was replaced by a hydrophilic amino acid, because in this case their functional or structural properties are more likely to be conserved.

So scoring matrices are devised to assign penalties accordingly to the above stated facts.

The following is an **Identity Matrix**:

	A	C	G	T
A	1	0	0	0
C	0	1	0	0
G	0	0	1	0
T	0	0	0	1

BLAST: A Database search technique uses a score of + 5 if the nucleotides are equal or a score of – 4 if they are not.

	A	C	G	T
A	+5	–4	–4	–4
C	–4	+5	–4	–4
G	–4	–4	+5	–4
T	–4	–4	–4	+5

There is another type of matrix which is called the **Transition Traversion Matrix** which gives a score of +1 if the nucleotide is similar and a score of –1 if a purine is replaced by a purine (i.e A or G) and also a score of –1 if a pyramidine is replaced by a pyrimidine (i.e C or T). If a purine is replaced by a pyrimidine or a pyrimidine is replaced by a purine then it gives a score of –5 as shown below:

	A	C	G	T
A	1	–5	–1	–5
C	– 5	1	–5	–1
G	–1	–5	1	–5
T	–5	–1	–5	1

Several scoring matrices can be devised based on the physical and chemical properties of an amino acid.

For example, pairing two amino acids which are charged may have a +ve score and pairing an amino acid which is polar with a hydrophobic may have a –ve score.

8.6 POINT ACCEPTED MUTATION MATRIX

A point accepted mutation matrix or (PAM) is a matrix in which each column and row represents one of the 20 standard amino acids.

The entry of a PAM matrix shows the likelihood of the amino acid in the row to be replaced by the amino acid of the column.

The first PAM matrix constructed like this is called PAM –1 and it shows one evolution. Other evolutions can be shown by constructing a series of PAM matrices which indicate the evolution of time.

The following is a PAM matrix:

	A	C	D	E	F	G	H	I	K	L	M	N	P	Q	R	S	T	V	W	Y
A	4																			
C	–1	6																		
D	–2	–1	2																	
E	–1	2	1	5																
F	–3	3	4	2	9															
G	–1	–2	1	0	1	3														
H	2	1	3	5	2	4	8													
I	1	–1	–2	0	1	–1	1	5												
K	1	–1	–2	2	2	3	–1	4	6											
L	1	4	5	6	3	4	5	6	8	10										
M	–1	2	3	–4	4	5	3	2	4	5	7									
N	–1	–2	3	4	2	5	3	6	3	5	3	9								
P	3	6	7	5	4	8	9	10	8	12	11	10	15							
Q	–1	2	–2	3	–3	2	4	1	–1	2	–1	1	1	5						
R	2	4	6	8	11	10	9	8	10	7	6	12	13	15	18					
S	2	3	6	17	14	11	15	10	9	7	18	15	13	12	11	20				
T	4	5	7	12	9	8	10	13	10	11	6	7	8	11	12	10	16			
V	2	3	4	–1	5	1	2	–1	6	1	7	6	5	3	4	2	9	10		
W	–1	2	–3	4	5	6	7	8	–1	2	7	3	5	–1	5	3	6	8	9	
Y	–1	2	–3	1	–1	4	3	2	–2	1	–2	1	3	4	2	1	4	2	1	5

The amino acids may be listed alphabetically or listed in groups.

The top triangular matrix would be similar to the lower triangular matrix because they represent the same substitution.

A PAMn matrix indicates that n mutations would have occurred per 100 amino acids. A commonly known PAM matrix is PAM250 which assumes that 250 mutation events would have occurred per 100 amino acids.

PAM-1, this unit is 1 substitution (or accepted point mutation) per 100 residues, or one PAM unit.

PAM-1 matrix is used to compare sequences that are closely related. PAM-1000 matrix are used to compare sequences at very distant relationships. The most commonly used is the PAM250.

A positive value in the PAM matrix entry indicates that it is a point accepted mutation and a negative entry indicates that it is more likely due to chance.

$$\text{Relative\_Mutation (i)} = \frac{\text{Mutation (i)} * \text{ freq (i)} * 100}{\text{Mutation\_Total} * 2} \qquad \text{eq. (1)}$$

$$\text{Mutation\_Probability (i, j)} = \frac{\text{Relative\_Mutation (i)} * \text{Substitution (i, j)}}{\text{Substitution (i)}} \qquad \text{eq. (2)}$$

$$\text{PAMn (i, j)} = \frac{\log \text{(Mutation\_Probability (i, j))}}{\text{freq (j)}} \qquad \text{eq. (3)}$$

Since we are taking the log in the above case it means positive value indicates the numerator is high and negative value indicates the denominator is high.

Now consider the following phylogenetic tree to construct the PAM matrix.

Suppose for above equations i = the amino acid I

j = the amino acid C

The following mutations have occurred in the evolution tree

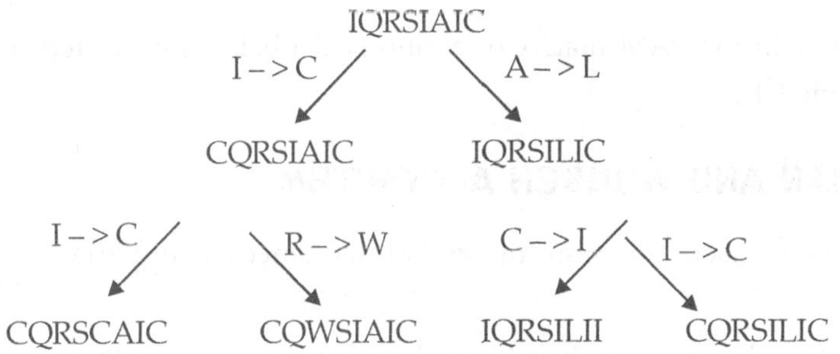

Mutation (i) = The number of mutations involving I = 4

freq (i) = The occurrence of total number of I divided by total number of amino acids.

- 17/56

= 0.3035

Mutation _Total = 6

The relative mutation is the number of times an amino acid is substituted by another amino acid and then this number is divided by the total number of mutation that would have affected the residue.

A scaling factor of 100 is used and By using eq (1)

$$\text{Relative-Mutation (i)} = \frac{4 * 0.3035 * 100}{12}$$

$$= 10.11$$

Here I has mutated 4 times.

Using eq (2)

$$\text{Mutation\_Probability (i, j)} = \frac{10.11 * 4}{4}$$

Where 4 in the numerator is the substitution of (i, j) in either direction i.e., I – > C or C – > I.

And 4 in the denominator is the substitution involving I which is again 4.

$$\text{Mutation\_Probablity} = 10.11$$

$$\text{PAMn}(i, j) = \frac{\log(10.11)}{\text{freq}(j)}$$

where freq (j) is the number of occurrences of C out of the total number of amino acids.

$$= 11 / 56$$

$$= 0.1964$$

$$\text{PAMn}(i, j) = \log(10.11/0.1964)$$

$$= \log(51.47)$$

$$= 1.711$$

This is the entry in the PAM matrix of amino acid i being substituted by amino acid j. In this case I substituted by C.

8.7 NEEDLEMAN AND WUNSCH ALGORITHM

This algorithm is used for aligning global sequences using the concept of dynamic programming.

If you think of a sequence which is 90 nucleotides long and another sequence which is 80 nucleotides long and you start comparing each possibility of 90 with each possibility of 80 then it would take enormous amount of time.

As the nucleotide sequence number increases it becomes impossible to keep a track of which is the best possible alignment.

To answer the above question, Needleman and Wunch were the first ones to propose an algorithm which constructs a matrix dynamically and finds the best alignment.

It assumes the following property:

+1 for a match

0 for a mismatch

–1 for a gap penalty

we take the following two sequences as an example:

TCGATTACGT

CATCGTAT

We plot a matrix of two sequences on the X and Y axis as follows:

		C	A	T	C	G	T	A	T
	0	–1	–2	–3	–4	–5	–6	–7	–8
T	–1								
C	–2								
G	–3								
A	–4								
T	–5								
T	–6								
A	–7								
C	–8								
G	–9								
T	–10								

The initial row and column of the matrix is filled with the multiples of gap penalties.

Now the rest of the matrix is filled by the following criterion:

1. Either we put a gap in the horizontal axis sequence
2. Or we put a gap in the vertical axis sequence
3. Or we align the base pairs, they may match or mismatch

The best alignment is calculated by following the path from the upper left corner of the matrix to the lowest right column in the matrix.

A horizontal move in the table represents a gap in the sequence along the left axis. A vertical move represents a gap in the sequence along the top axis and a diagonal move represents an alignment of the nucleotides from each sequence.

The matrix is filled with the matrix position (2, 2)

This position can be filled in 3 ways:

1. Introduce a gap in the top sequence
2. Introduce a gap in the left sequence
3. No gap, align the base pairs

To do this if we

1. Introduce a gap in the top sequence and take the value from the top cell which is (2, 1), it is –1 and add a gap to it, it becomes –2.
2. Introduce a gap in the left sequence and take the value from the left cell which is (1, 2), it is also –1, add a gap to it –1, which becomes –2.

3. Take the value from the diagonal element above (1, 1) and add either a match bonus (+1) or a mismatch bonus (0). Here T and C do not match so mismatch bonus is 0.

Out of the above 3 values take the maximum value and fill the cell with that value. i.e., –2, –2 0. So 0 is maximum so we fill 0 in the matrix (2, 2) position.

Similarly fill all the cells of the matrix:

		C	A	T	C	G	T	A	T
	0	–1	–2	–3	–4	–5	–6	–7	–8
T	–1	0	–1	–1	–2	–3	–4	–5	–6
C	–2	0	0	–1	0	–1	–2	–3	–4
G	–3	–1	0	0	–1	+1	0	–1	–2
A	–4	–2	0	0	0	0	+1	+1	0
T	–5	–3	–1	+1	0	0	+1	+1	+2
T	–6	–4	–2	0	+1	0	+1	+1	+2
A	–7	–3	–3	–1	0	+1	0	+2	+1
C	–8	–4	–3	–2	0	0	+1	+1	+2
G	–9	–5	–4	–3	–1	+1	0	+1	+1
T	–10	–6	–5	–3	–2	0	+2	+1	+2

Now to fill the position (2, 3), we take value from top cell which is –2, add gap penalty which is –1 so it becomes –3. We take the value from the left cell (0), add gap penalty which is –1 so it becomes –1. We take the diagonal value (–1) and add a mismatch (0) so again it becomes –1, so we can fill the cell with a value of –1.

- C - A T C G T A T

T C G A T T A C G T

Now the alignment is done as follows:

We start from the bottom right column and see from the left, top and diagonal element that how that number would have been produced.

For the cell (8, 10), it is +2, now from the top and left they would have given 0 because a gap penalty to +1 would have resulted in 0 in both cases, and a match to the diagonal element +1 would have resulted in +2. So we traverse backward to the position of the diagonal column with an arrow.

Now we see how cell no (7, 9) would have resulted in +1. The top and left cell would have given 1 – 1 = 0, and 0 – 1 = –1 respectively because of the gap penalties and a mismatch with the diagonal column would have given + 1 + 0 = +1. So +1 is the highest and again we traverse to the diagonal column and similarly reach to the top.

We find the following arrow sequence:

↑ ↖ ↑ ↖ ↖ ↖ ↖ ↖ ↖ ↖

8 diagonal arrows means matching of base pairs and 2 vertical arrows means gap in the top axis, since there are no horizontal arrows so there are no gaps in the left axis.

So the alignment becomes according to the arrow sequence:

- C - A T C G T A T

T C G A T T A C G T

Here only 3 base pairs match, but this is the best possible alignment that we can get out of the two sequences.

8.8 SEMI GLOBAL ALIGNMENT

The Needleman and Wunsch algorithm is an example of the global alignment. It does not distinguish between gaps that are at the beginning or end of the sequences. Generally the gaps which are at the beginning or end are known as terminal gaps and have less importance than those gaps which are in the middle. Now you will ask the question why is this so. This is because terminal gaps are an indication that here the sequences would have been broken or the data was incomplete. Hence they are of less significance to us.

The concept of semi global alignment takes this into account and does not provide a gap penalty for those gaps which are terminal.

A vertical move in the previous algorithm was a gap in the top most axis and a horizontal move was a gap in the left axis.

If we want to give no penalty to our top most axis at the beginning of the sequence or for terminal gaps then we can modify our Needleman and Wunsch algorithm by making all zeros in the first column. This will allow our first sequence to not give a penalty for the beginning gaps. Similarly if we make our first row equal to all zeros then gaps in the second sequence at the beginning will be without penalty.

Now this was the case when we have taken into account the gaps at the beginning of both the sequences.

What do we do about the gaps that will be at the end of both the sequences. Remember again a horizontal or a vertical move indicate a gap in the left axis and the top axis respectively. So gaps in the last row which are all horizontal indicate end gaps in the second sequence (left sequence) and we will make them free without any gap penalty. Similarly vertical moves along the last column be left free without penalty will indicate end gaps in the first sequence (top sequence).

An example of a semi global alignment is:

Consider two sequences:

TCATAGCTAT

TCAT - - - - - -

		T	C	A	T	A	G	C	T	A	T
	0	0	0	0	0	0	0	0	0	0	0
T	0	1	0	0	1	0	0	0	1	0	1
C	0	0	2	0	0	0	0	1	0	0	0
A	0	0	1	3	0	1	0	0	0	1	0
T	0	0	0	2	4	4	4	4	4	4	4

So here the horizontal arrows which indicate the terminal gaps are not given any importance.

8.9 THE SMITH-WATERMAN ALGORITHM (LOCAL ALIGNMENT)

This relies on local alignment rather than semi global alignment. In the semi global alignment we give a gap penalty of –1, a match bonus of +1 and a mismatch score of 0.

Even if we allow zeros for terminal gaps, the semi global alignment may lead to weird alignment scores because of the mismatches involved.

This leads us to step ahead of semi global alignment which looks for local alignments and is known as the Smith-Waterman Algorithm.

For local alignment, the global alignment algorithm remains the same, only another option is added that if we have a score below 0 out of the three options, we make it a 0 instead of taking a –ve value.

Then we make our way backwards from the maximum score to a value where we get a zero.

The local alignment ignores any mismatches or gaps before or after the matching region.

Suppose we have two sequences:

CATGCTTAACGT

GCATTAAC

		C	A	T	G	C	T	T	A	A	C	G	T
	0	0	0	0	0	0	0	0	0	0	0	0	0
G	0	0	0	0	1	0	0	0	0	0	0	1	0
C	0	1	0	0	0	2	1	0	0	0	1	0	0
A	0	0	2	0	0	0	0	0	1	1	0	0	0
T	0	0	1	3	2	1	1	1	0	0	0	0	1
T	0	0	0	2	1	0	1	2	1	0	0	0	1
A	0	0	1	1	0	0	0	0	3	2	1	0	0
A	0	0	1	0	0	0	0	0	2	4	3	2	1
C	0	1	0	0	0	1	0	0	1	0	5	4	3

We obtain the following local alignment

TTAAC

TTAAC

Here the maximum value is 5 and we make our way backwards till zero.

8.10 DATABASE SEARCHES

In database searches, we have a sequence of an unidentified gene and we try to align this unidentified gene with millions of sequences that are present in our Bioinformatics database such as GenBank to find out its functional role or whether it matches with some other gene whose role has already been defined. To find out any expression in which it is involved or how it is being regulated or whether it shows any kind of relationship with any other gene.

In finding the sequences that align well, many index based searches are used because it becomes too difficult to find searches that align well with a huge number of sequences. Therefore some kinds of tools are required to make the job easier.

8.11 BLAST

BLAST algorithm looks for sequences which have the maximum ungapped local alignments. From a larger query sequences, a number of subsequences are searched. The BLAST algorithm may be used for both kind of searches, either protein sequences or nucleotide sequences.

There are many variations of the BLAST algorithm. One is BLASTP which searches for protein sequences. Others are BLASTN and BLASTX, where BLASTN searches for nucleotide sequences and BLASTX translates nucleotides to proteins before searching.

To illustrate one of the BLAST algorithm, we will take the BLASTP algorithm and work it through an example.

The BALSTP algorithm first breaks down a large query sequence into equal sized words of small lengths say 4 letters or 5 letters. It then makes different words by first starting with the first letter in the subsequence and making a word of 4 letters (say). Then it shifts one letter to the right and makes another word of 4 letters. The next word is formed by shifting 2 letters and then forming a 4 letter word and in this manner a number of possible words are made which are to be searched.

The sequence is then searched where these words can be matched. Once a word is matched it is extended in both directions. So that more letters can be matched if possible. This is confirmed till the matching score does not fall below a given mark. And the local alignments are ungapped so no gaps are allowed, simple matches have to take place until the score keeps on increasing or does not fall below the minimum threshold mark.

For Example take the query sequence:

IQRSIALCTVP

The possible number of words that can be made are:

1. IQRS
2. QRSI (shifting 1 letter to the right)
3. RSIA (shifting 2 letters to the right)
4. SIAL (shifting 3 letters to the right)

And so on

The second step is to match the words with the database sequence.

Suppose the database sequence is:

ADLYFLMCQTYQRSIGTCTMLMVQ....

The third step is to extend in both directions until a minimum threshold mark is reached.

→
QRSIALCT

ADLYFLMCQTYQRSIGTCTMLMVQ.....

The threshold mark is to be chosen with great care depending upon the type of search you are looking for.

8.12 FASTA

The difference between FASTA and BLAST is that BLAST performs ungapped local alignments and FASTA performs alignments which allow the gaps. Also FASTA is more helpful when there is a repetition of the amino acids.

First the sequence with which comparison has to be made is taken and a table is constructed which gives the positions of the various amino acids in the sequence. If an amino acid occurs twice then 2 entries are made for that particular amino acid.

For example take the 20 amino acids and prepare their chart.

letter	A	C	D	E	F	G	H	I	K	L	M	N	P	Q	R	S	T	V	W	Y
Position																				

If we take the query sequence as:

IKLCNPQIDECTVW

letter	A	C	D	E	F	G	H	I	K	L	M	N	P	Q	R	S	T	V	W	Y
Position		4	9	10				1	2	3		5	6	7			12	13	14	
		11						8												

As you can see C occurs at 4 and 11th position, so it is shown that way in the chart.

This sequence may be compared to the target sequence which is as follows: NPQIDECLYTM

Next we prepare a chart of the target sequence with respect to the positions of the query sequence by finding at which position the amino acid is in the query sequence and subtracting the position of the target sequence amino acid with the query sequence amino acid.

position	1	2	3	4	5	6	7	8	9	10	11
Target	N	P	Q	I	D	E	C	L	Y	T	M
	4	4	4	4	4	4	–3	–5	5	2	
							4				

Here N is found at position 5 in query sequence and 1 in target sequence so 5 – 1 = 4.

Similarly P is found at position 6 in query sequence and 2 in target sequence so 6 – 2 = 4.

C is found at two places, 4th position in query sequence and 7 in target sequence so 4 – 7 = –3.

Also it is found at 11th position in query sequence and 7th in target sequence so 11 – 7 = 4.

Since M amino acid is not included in the query sequence so it is left empty in the target sequence table. Now you can see that most entries show 4 in the target sequence table so we can shift by a value of 4 and get our results.

IKLCNPQIDECTVW NPQIDECLYTM

So the above is the best possible match.

8.13 CONNECTIONS TO CLINICAL DATA

In Bioinformatics, connections to clinical data means that given the clinical history of a patient regarding any of the disease, how his genetic make up matches to that disease. Whether we can predict that his offspring will have the same kind of disease, or whether can we make any drug which will cure his disease according to his genetic make up. These are the issues which are related to the connections to the clinical data regarding a person.

For doing this, we first have to look at the data and make information out of it and then progress it to metadata which is data about the data.

In order to do so we require in this process a datawarehouse, a number of data repositories, a number of datamarts, data dictionaries, clinical database records and genomic databases, online medicine databases such as MeSH and online literature reference databases such as PubMed.

Data are collected from a variety of sources. The patients data is fed into a microarray sequence data from which gene expression is done to find its medical relevance and sent to a data mart. The patients disease is examined and all clinical reports are again fed into another data mart. Then these 2 data marts along with public private databases such as Gen Bank PubMed and MeSH are used together in connection with taxonomic data and data dictionary to be fed into a dtawarehouse from which various patterns can be extracted and data can be mined.

Now the patterns that are mined from the datawarehouse focus on how to identify the genetic basis of the particular disease that the patient has. Then genomic sequences of other animals such as rats, fruit flies are also compared to find the particular gene which is responsible for the disorder.

From the genetic studies, may be there is a possibility to find the hormones or neurotransmitters that are responsible for a particular disorder.

From the medical literature, the scientists may propose a new drug that moderates the hormones to recover from the disease.

Then its connection to other diseases can also be found out and the symptoms of those iseases can be notified.

From these databases is created an EMR report of the patient which is the Electronic Medical Record (EMR), which is a modern approach to create a patient's clinical status which can be stored in any of the central databases for access to all.

Then with the help of the data dictionary these reports can be converted into a single format which is understood by all.

The advantage of a datamart is to store together the genomic clinical and psychological data together which can be visualized and provide information that can be of great importance in learning how, why and what things are taking place pertaining to that disease.

Pharmacogenic data is also available simultaneously for doing any kind of qualitative or quantitative analysis or simulations or structural analysis etc.

8.14 MULTIPLE SEQUENCE ALIGNMENT

In multiple sequence alignment, a number of sequences are taken and their gaps are aligned in such a way that maximum possible alignment takes place between the gaps of all sequences in comparison to taking 2 sequences and doing pairwise comparison.

The complexity involved in calculating the multiple sequence alignment is far greater and thus algorithms discussed for pairwise sequence alignment do not work well for this.

One of the algorithm used for multiple sequences is the CLUSTAL algorithm which first aligns the sequences which are closely related and then expands by choosing sequences which are not closely related and fixing their positions in the already aligned sequences one by one.

A variation of CLUSTAL is the CLUSTALW in which weights are assigned to find out how closely the sequences are related or how distantly the sequences are related.

Making a matrix out of multiple sequence alignments is very difficult as the number of dimensions increase and to keep a track of of that is even more difficult and also the complexity increases tremendously.

Like pairwise sequences, multiple sequence alignments are meant for amino acid sequences or DNA sequences and the motive is to find out the structural, functional similarities among a varying number of sequences taken from a variety of animals and to find out anything that is of significant importance that would lead to discovery of new knowledge about protein or the gene.

The upcoming methods apart from the CLUSTAL algorithm are the ones which use genetic algorithms and Hidden Markov Models as their base.

In a genetic algorithm first the population is taken, the best parents are chosen according to the values of a fitness function. The parents are then crossed over to make offsprings, these offsprings form the new generation and then certain mutations can be done in these offsprings. The generated solution is fed back replacing the initial population and then the process is repeated until an optimal solution is derived.

In the context of Bioinformatics, first a random number of gap insertions and deletions are done with the alignment score being taken the value of the fitness function. The sequences which give the best fitness value are allowed to crossover and produce offsprings, the new generation offsprings can have certain mutations done on a small number of amino acids like replacing one amino acid with another amino acid. After the mutations the newly generated sequences are allowed to cross over again, again the ones having the best fitness value are selected and they create new patterns. In this way the cycle is repeated until a multiple alignment solution which is optimal is achieved.

8.15 SECONDARY DATABASES

The secondary databases are also referred to as pattern databases because they contain the data after analysis of the primary databases.

From primary databases it is difficult to do analysis and hence there is the requirement of secondary databases.

It is important to do analysis separately mainly because the data is repetitive, may have noise, the large size of it would be difficult to handle. There may be ambiguity and the data may have been cut at ends.

Also species of the same gene family show the similarity which may not be detected by primary databases, so secondary databases help in finding relationships between superfamily, family, subfamily and results of significant importance are revealed through secondary databases.

How Secondary databases are formed?

The multiple sequences where conserved regions from different species were aligned were known as motifs. These motifs led to the insight that it has some significant property either in terms of a functional role or a structural role and this motif led to a unique feature of the gene family belonging to different species in that gene family. This may also be known as a kind of signature of the gene family which is unique in nature and tells us something of significant importance about that gene family.

So different secondary database are made based on the method used to get to the family signature after analyzing the motifs and the different types of secondary databases are described in the following sections.

8.16 REGULAR EXPRESSIONS

Regular expressions show that how conserved regions in a motif can be expressed in the form of an expression. Here we tell at what positions which amino acids are present in all the sequences. At what position which alternate amino acids are allowed, at what positions which amino acids are completely disallowed. What other things are allowed at other positions which are left out.

For example, take the following multiple sequences which have been aligned:

IVMSALPFPGCD
ILNSBFCQPGGD
IVPSDEKFPCGD
IVQSEDLBPGCD

The regular expression for the above sequence would be:

I – [LV] – {FC} – S – x3 – [FQB] – P – [GC]2 – D

Here in all we have 12 positions, out of which position no 1, 4, 9, and 12 are completely conserved i.e. all of them show the same amino acid at that particular position. In this case I, S, P and D respectively are the amino acids.

In the regular expression the first possible amino acid is given separated by a hyphen (–). The next amino acid is given in a square bracket. The square brackets indicate that any one of the amino acids in the square brackets is allowed, so here [LV] says that either L can exist or V can exist. The third position is indicated by a curly bracket, the curly bracket says that none of the amino acids in the given brackets is allowed, any other amino acid can exist, so we have {FC}, i.e neither F nor C is allowed at 3rd position any other amino acid can exist. Then again the 4th position is given by S so it is present at the same position in all the sequences. Then next x3 indicates that next 3 position can be filled with any amino acid without any restriction. The next position is given by [FQB] in square brackets, so any one of the 3 amino acids either F or Q or B can be allowed at that position. Next amino acid P is completely conserved. Then we have [GC]2 which means that these possible amino acids can occur in a combination of 2 in any order, therefore there is a number 2 beside it. Then the last position D is completely conserved.

The problem with the regular expression is that they are very strict, they impose that these amino acids should be present or not present. If we have one amino acid extra in a new sequence or an amino acid less in another sequence then it does not cater for it.

Also another problem is that this regular expression is based on a single motif. Suppose there are other motifs which can show better conserved regions or have come into picture later, but now the family is already known by a previous regular expression created from a previous motif. So it does not take into account more number of motifs, it depends only on a single motif.

The regular expression entries occupy the position in the PROSITE secondary database.

8.17 RULES

Regular expressions which were formed from a motif were generally 10-20 amino acids in length and they tell about the structural or functional role of a protein in that gene family.

Rules on the other hand are 3-4 amino acids in length and tell about some sites in particular such as sugar attachment sites or phosphate attachment sites etc. They cannot be used for any analysis of a structural or functional role of a protein, they can simply tell about the beginning of some site.

Example of a Rule:

Sugar Attachment Site: N – {P} – [ST] – {P}

Now this is only 4 positions in length, the first alphabet would be N, the second can be anything apart from P, the third may be S or T and the 4th can be anything apart from P.

The rules entries occupy the position in the PROSITE secondary database.

8.18 FUZZY REGULAR EXPRESSIONS

In fuzzy regular expressions, some amount of fuzziness or tolerance is introduced into the regular expressions so that amino acids having similar properties are allowed at that position.

For example we have the following amino acids with their properties:

Non Polar – F, M, A, V

Polar – T, S, Y, N, Q

Hydrophobic – L, I, W

Hydrophilic – R, K, H, P

Then our regular expression can become:

[FMAV] – C – [TSYNQ] – D-[LIW] –[RKPH] – Q

So the first position allows any of the non polar amino acids either F, M, A, or V. Then the 2nd position is conserved by amino acid C, then 3rd position allows any of the polar amino acids like T, S, Y, N, Q, then the 4th position is conserved by amino acid D, then 5th position allows hydrophobic amino acids L, I, W and 6th position allows hydrophilic amino acids R, K, P and H. And the last position is conserved by amino acid Q.

In this way we have introduced fuzziness and given enough choices at one position.

The advantage of this is that the strictness is finished and lot of amino acids will be allowed to match now.

The disadvantage is that by allowing so many amino acids at one place we introduce too many matches and they may lead to misleading results as the amino acids will not be properly conserved.

The fuzzy regular expressions fill the entries of the BLOCKS and PRINTS secondary databases.

8.19 FINGERPRINTS

Fingerprints take into account several motifs that are taken from the multiple sequence alignment. The proper way would be to take the conserved regions from all of these motifs and make the signature.

Fingerprints is a technique in which the matrix is built of the number of times a amino acid has occurred at one position.

A matrix containing all the possible amino acids is labeled across the X axis and along the Y axis are labeled the number of positions in a motif. The matrix cells are filled with how many times has that amino acid occurred at that position.

First consider the following multiple sequences:

T A G S P Q L R T A H K
T G C P Q R L M B E D K
T N B S L M L D F G C K
T A M S R T L M C G H K
T A L S R M L D F G C K
T G C P M M L R C A H K
T A D S M L L D F A H K

Now consider the following table of amino acids:

0	0	0	0	0	0	0	0	0	0	0	0	0	0	0	0	0	7	0	0	0	0	0
4	0	0	0	0	0	2	0	0	0	0	0	1	0	0	0	0	0	0	0	0	0	0
0	1	2	1	0	0	1	0	0	0	1	1	0	0	0	0	0	0	0	0	0	0	0
0	0	0	0	0	0	0	0	0	0	0	0	0	2	0	0	5	0	0	0	0	0	0
0	0	0	0	0	0	0	0	0	0	1	2	0	1	1	2	0	0	0	0	0	0	0
0	0	0	0	0	0	0	0	0	0	1	3	0	0	1	1	0	1	0	0	0	0	0
0	0	0	0	0	0	0	0	0	0	7	0	0	0	0	0	0	0	0	0	0	0	0
0	0	0	3	0	0	0	0	0	0	0	2	0	0	0	2	0	0	0	0	0	0	0
0	1	2	0	0	3	0	0	0	0	0	0	0	0	0	0	0	1	0	0	0	0	0
3	0	0	0	1	0	3	0	0	0	0	0	0	0	0	0	0	0	0	0	0	0	0
0	0	2	1	0	0	0	4	0	0	0	0	0	0	0	0	0	0	0	0	0	0	0
0	0	0	0	0	0	0	0	0	7	0	0	0	0	0	0	0	0	0	0	0	0	0

There are 7 sequences in the above example with each 12 amino acids long. So the table has 12 entries.

It clearly shows from the matrix that T, L and K amino acids are completely conserved by simply looking at the matrix and finding 7 written in the cells corresponding to it.

The method is not very useful as the distant relatives belonging to the same gene family cannot be found easily. The fingerprint technique is used for the PRINTS secondary database.

The fingerprints technique becomes helpful when more number of séances are added to the existing number of sequences. When the database is run a number of times repeatedly to add additional sequences to the family motifs then clearly the ones which are conserved are shown in a better way.

For example taking the previous example, after a number of iterations the matrix values change to:

A	B	C	D	E	F	G	H	I	K	L	M	N	P	Q	R	S	T	V	W	X	Y	Z
0	0	0	0	0	0	3	0	0	0	0	0	0	0	0	0	0	45	0	0	0	0	0
21	0	0	0	0	0	5	0	0	0	0	0	3	0	0	0	0	0	0	0	0	0	0
0	5	10	12	0	0	3	0	0	0	7	6	0	0	0	0	0	0	0	0	0	0	0
0	0	0	0	0	0	0	0	0	0	0	0	0	10	0	0	25	0	0	0	0	0	0
0	0	0	0	0	0	0	0	0	0	5	12	0	18	11	12	0	0	0	0	0	0	0
0	0	0	0	0	0	0	0	0	0	10	21	0	0	10	5	0	4	0	0	0	0	0
0	0	0	0	0	0	0	0	0	39	0	0	0	0	0	0	0	0	0	0	0	0	0
0	0	0	18	0	0	0	0	0	0	0	14	0	0	0	20	0	0	0	0	0	0	0
0	5	15	0	0	3	0	0	0	0	0	0	0	0	0	0	0	1	0	0	0	0	0
30	0	0	0	1	0	3	0	0	0	0	0	0	0	0	0	0	0	0	0	0	0	0
0	0	11	1	0	0	0	28	0	0	0	0	0	0	0	0	0	0	0	0	0	0	0
0	0	0	0	0	0	0	0	0	35	0	0	0	0	0	0	0	0	0	0	0	0	0

Now the number of sequences has increased to 45 instead of 7 and the ones being conserved having values above 20 can easily be shown. After the end of the iteration only position 1 is completely conserved rest values have deviated.

8.20 BLOCKS

In fingerprint scheme, the weights are not used whereas in Blocks weights are used. This gives entries to the BLOCKS secondary database.

In this the conserved motifs are called blocks and are found by searching for spaced amino acids triplets (F – - C – G) etc. The score is calculated by BLOSUM 62 substitution matrix and is verified by a motif finding algorithm which searches for the highest score of the blocks that occur in a particular order. When blocks found by both methods match then they are entered in the database.

Sequences from various organisms are put together according to the weights assigned to them. This is called a cluster. The ones with smaller weights are closely related and ones with higher weights are distantly related. The weight of 100 specifies highest dissimilarity.

For example consider the entries from the BLOCKS database:

CCKR_MOUSE	(315)	APMVLMQRSTBDE	5
CCKR_FRUITFLY	(382)	APMVLMQRSTBDE	5
OPSD_HUMAN	(215)	MNPADEFQRST	14
OPSD_PIG	(254)	MNPADEFQRST	14
OPSD_RAT	(238)	MNPADEFQRST	14
OPSD_FRUITFLY	(203)	MNPADEFQRST	14
YN84_CAEEL	(313)	ADMVPSTRDAEQ	100

In this the entries from the SWISS_PROT are taken where the ID identifies the organism. The next number in brackets identifies the starting sequences, then comes the conserved sequence itself and then comes the weights (5, 14 and 100) in this case.

The organism sequences are clustered together according to their weights.

Care should be taken about a Blocks entries because high scores may lead to less significant meanings and low scores may be having more significant biological information hidden in it.

8.21 PROFILES

Profiles is a technique where sequences are cleaned and scoring tables are made to form profiles.

Profiles give an insight that

1. Which amino acids will be allowed at which position
2. Which amino acids are completely conserved.
3. Which positions can accept insertion to take place
4. Which can tolerate deletions

The scores are given in the form of penalties which may vary for insertions and deletions.

An I indicates an insertion

An M indicates a match

A D indicates a deletion

An example profile entry contains:

/I : SY = 'ch1'; attributes

/I : SY = 'ch1'; attributes

/M : SY = 'ch2'; attributes

Where ch1 is the character representing an insert position in the parent sequence and ch2 represents a match position in the parent sequence.

The attribute list contains values to be assigned to different position related scores which may include insertion or match or deletion scores.

The blocks are conserved by assigning values of penalties in the first place. These values of penalties can vary.

They are of 4 types:

1. MI – Match Insert transition score
2. I – Insert extension score
3. MD – Match Delete transition score
4. D – Delete extension score.

For example we can give large penalties of each one of them like

MI = –20, I = –5, MD = –20, D = –5

And we can also assign lesser penalties in the form:

MI = 0, I = –1, MD = 0, D = –1

PROSITE is the secondary database where profile entries are stored.

For example an entry in the PROSITE database is as follows:

/DEFAULT : MI = –20, I = –5, MD = –20, D = –5

/M : SY = 'K' ; M = –6, –1, –2, –4, –3, –2, –1 0;

/M : SY = 'D' ; M = –7, –3, –1, –2, –6, –4, –1;

/I : MI = 0, I = –1, MD = 0, D = –1

/M : SY = 'A'; M = 0, –1, 1, –2, 0, 1, –1;

/M : SY = 'P'; M = –3, –4, –1, 0, –1 , –2;

Here we have shown two conserved blocks, one is the default block.

8.22 HIDDEN MARKOV MODELS

A variation to store sequences relies in the form of a Hidden Markov Model.

A Hidden Markov Model is a probabilistic model that has a number of interconnecting states. They are chains of match, insert or delete states connecting one another.

These chains hide in them the method in which sequences have been conserved within alignments.

A match state indicates that the amino acid at this place is conserved. An insert state indicates that an insertion can be done relative to the matched amino acid and a delete state indicates that a deletion can be done relative to the matched amino acid.

HMM is built from multiple sequence alignments by specifying that every position of the sequence has either a match, a delete or an insert state.

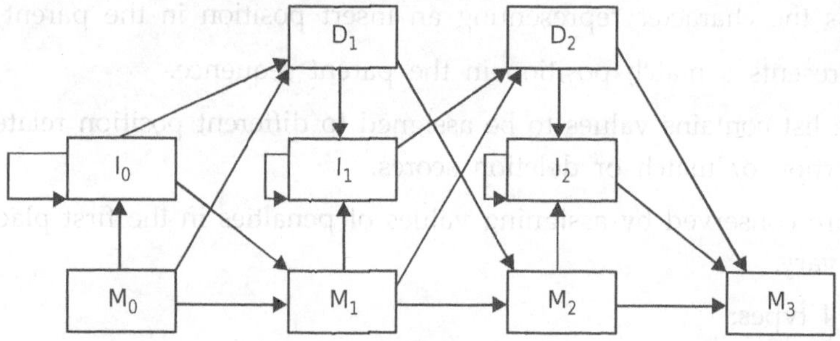

The above model says that first you start by either a match (M_0) or an insert (I_0), from that you can either go to a delete position (D_1) or an insert again (I_1) or to the next match (M_1). From the first match (M_0) you can either go to the delete (D_1) or to the second match (M_1) and in this way you can encode how sequences are conserved.

The HMM are used to fill the entries of Pfam secondary database.

Unit **9**

IT IN BIOINFORMATICS

9.1 PERL BASICS

Perl is a programming language which is most popular for the Bioinformatics data. It is easy and simple to use and is particularly helpful in handling the data used in bioinformatics.

In the field of Bioinformatics you would require to change the format of the data to keep it consistent with data collected from other sources. So the software which most suits these needs are available in the Perl language.

As most of you would be knowing that the computer understands the machine language that is binary code, only 0's and 1's. That is why the user finds it difficult to write the code in the form of 0's and 1's and writes in a high level language such as C. C++ which the user clearly understands since it is a language which uses English words and expressions to form sentences.

After this the complier translates this high level language into machine language and the machine language is the executable file which is normally given the extension of ".exe".

The Perl language does not require a compiler. It requires a Perl interpreter to run the Perl script. The Perl script is nothing but similar to your high level language in which a set of Perl commands are given which are to be executed one after another.

The Perl script takes an extension of ".pl". this Perl script can be run on Unix and Windows operating systems. Perl script is similar to Unix shell scripts where Unix commands are written in the shell scripts and run on the Unix operating system.

9.1.1 Variables

Like all programming languages, there is a concept of a variable, which holds the values in some storage space that is allocated to it in the memory area.

A name of the Perl variable like most other languages can be formed by using alphabets, numbers, underscore (_) and a hyphen (-).

Like Unix shell scripts, a variable is generally referred by putting a $ sign in front of the variable name such as $x where x is the variable name.

A variable in Perl behaves like any variable in other programming languages in which you can vary the value of the variable ant time.

The values to a variable are assigned by using the assignment (=) operator.

For example,

We can assign a value

```
$x = 2;
$y = 3;
$z = $x + $y;
```

Here $x, $y and $z are 3 variables. First $x and $y are assigned the values of 2 and 3 respectively and $z contains the sum of the values in $x and $y and therefore $z will contain the value 5. To see the value of $z, we can write the statement

```
print("value of z is $z\n");
```

This will print value of z is 5, like in C language "\n" is used to go to the next line, here also "\n" is used to go to the next line and "print" is a function used to display the value on the screen.

In Perl, it is not required to declare a variable, like in many other languages, you are required to declare variables in which you state what type of value the variable is going to hold. In Perl the variable can take kind of value that you assign to it and that will be its type, and you can do this in the part of the program without any specific declaration at the beginning of the program.

We can assign different values to variables like

```
$x = 2;
$y = 3.14;
$z = "hello";
```

So here x is an integer type variable, y is a real type variable and z is a string variable.

The Operators used in Perl language are as follows in order of Precedence:

+, –, * , /	->	For addition, subtraction, multiplication and division respectively.
%	->	For remainder
++ , --	->	Increment and Decrement, like in C language will increment the value of the variable by 1 and decrement will decrease the value of the variable by 1 and store the new value back into the same variable.
+=, -=, *=, /=	->	Shortcuts in for 1 assignment cum algebraic operation. Just as in C language. $x += 2; is equivalent to $x = $x +2;
**	->	Exponentiation $x**2 means $x = $x * $x; $x**3 means $x = $x * $x * $x;

. -> (dot) string concatenation if $x = "hello" and $y = "world" then $x.$y= "helloworld"

X -> string repetition $x = "my" X 3will make the variable x's value as $x = "mymymy"

9.1.2 Statement

A statement in Perl is similar to the statement in C language, which is terminated by a semicolon after writing the expression.

For Example:

```
$x = $y - $z * $p + $q;
```

Now here the precedence of operators will be used to solve the expression and the above is a statement in Perl language.

9.1.3 Block

Just like in C language any block is given by curly brackets {}. Here also more than one statement which is to be executed under one condition or the opening or closing of a function etc will be done by curly brackets { }.

9.1.4 IF ELSE Construct

Just like in C language, we can execute some part of the code based on some condition and not execute the other part of the code. This can be done with the help of if else constructs, whose variants are:

1. Single IF statement
2. IF ELSE statement
3. IF ELSIF statement

For example,

1.

```
$x = $report;
if ( $x < 10 )
{
 print(" The value is less than 10");
}
```

2.

```
if ( $x < 10 )
{
print(" The value is less than 10");
}
```

```
else
{
print("The value is greater than or equal to 10\n");
}
3.
if ( $x < 10)
{
print(" The value is less than 10");
}
elsif ( $x == 10)
print(" The value is equal to 10");
}
else
print(" The value is greater than 10");
}
```

The first example will print value is less than 10 if the $x variable has a value which is less than 10 otherwise it will not print anything.

The second example will print "the value less than 10" if the variable has a value less than 10 and exit. Otherwise if the if condition is false, it will print "the value is greater than or equal to 10" and exit.

The third example will print "the value is less than 10" if the value is less than 10 otherwise it will check the next condition whether it is equal to 10, if it is equal to 10 then it will print "the value is equal to 10" and exit. If the elsif condition is also false then it will print "the value is greater than 10" and exit.

The Relational and Logical Operators according to Precedence in Perl are as follows:

<, >, <=, >=	->	In case of numeric variables less than, greater than, less than or equal to and greater than or equal to respectively.
lt, gt, le, ge	->	In case of string variables less than, greater than, less than or equal to and greater than or equal to respectively.
==, !=	->	In case of numeric variables is equal to and not equal to respectively.
eq, ne	->	In case of string variables equal to and not equal to.
&&, ||	->	&& - According to truth table of AND operator || - According to truth table of OR operator
!	->	According to truth table of NOT operator.

Truth table and

X	Y	Answer
True	True	True
True	False	False
False	True	False
False	False	False

Truth table or

X	Y	Answer
True	True	True
True	False	True
False	True	True
False	False	False

Truth table not

X	Answer
True	False
False	True

The conditions can be combined by the && and || operators as follows:

```
if ( ( $x < 1) && ($y > 1) )
```

will return true or false, a value according to the true or false values returned by 2 conditions ($x < 1) and ($y > 1) and then use the truth table of AND to get the final value and then the if part may or may not be executed.

9.1.5 Loops

There are 3 kinds of loops in Perl language:

1. While Loop
2. For Loop
3. Foreach loop

A loop is block of statements which are executed repeatedly until the loop regulating condition becomes false.

Every loop has a condition at the beginning according to which the loop operates.

1. For example taking the while loop

```
$x = 5;
while ($x < 10)
{
  print ("x is equal to $x \n");
 $x = $x + 1;
}
```

Here at the beginning before the while loop begins, we have assigned the value 5 to the $x variable.

The while loop is regulated by the condition

($x < 10)

Until this condition is true, the statements inside the block of the while loop which is indicated by curly brackets { } will execute repeatedly. The first time the condition is checked whether the value of x is less than 10 or not. So the value of x is 5 and is less than 10, so it enters the loop, now it executes the first statement of the loop which is

```
print ("x is equal to $x \n");
```

It will print that on the screen and then go to the second statement which is

$x = $x + 1;

So it will increase the value of x from 5 to 6.

Now the loop's block ends with a curly bracket "}", so the control goes back to the beginning of the loop where the while statement is and checks the condition again, this time it will check

6 < 10, since the value of x has changed from 5 to 6, now again the condition is true, so it will enter the loop and proceed again in the similar manner.

Now on the screen it will print the output as:

x is equal to 5

x is equal to 6

x is equal to 7

x is equal to 8

x is equal to 9

Now the loop terminates when the value of $x has become 10, because when it will compare

10 < 10, it will be false because 10 is equal to 10, so it will not go inside the loop and come out of the while loop.

2. The For loop's syntax is the same as in C language

```
for ( $x = 5; $x< 10 ; $x ++)
{
 print( " x is equal to $x \n");
}
```

In the for loop the first expression is called the initialization statement, which is ended by the semicolon, the next expression is the regulating condition of the loop which is again ended by a semicolon and the 3rd is the step statement.

This is exactly similar to the while loop just described and it will print the same result as the while loop also.

First we are initializing the value of the variable $x to 5, then it checks the condition whether

$x < 10 or not , it is so, so it goes inside the loop straight away and will not go to the step statement now. Now it will execute the statement

```
print ("x is equal to $x \n');
```

then before coming out of the for loop it will now execute the step statement which is "x++", which is same as $x = $x + 1. So it will increment the value of x by 1 and x now has a value 6. Now it will not go to the initializing statement, it will just check the condition $x < 10, so 6 < 10, it is true, so it will enter the loop and execute the print statement. Now it will execute the step statement and next check the condition again. The loop will terminate when $x will have a value of 10 and the condition $x < 10 becomes false because x is equal to 10 and not less than 10. So again it will print the same output as while loop did.

3. The third type of loop is the foreach loop in which for every value given in the brackets of the foreach statement, the variable takes that value and executes accordingly.

Example,

```
foreach $var (2, 6, 3, 4, 8)
{
print("var variable has value equal to $var \n");
}
```

This will print the output on the screen as follows:

var variable has value equal to 2

var variable has value equal to 6

var variable has value equal to 3

var variable has value equal to 4

var variable has value equal to 8

9.1.6 Comment

The comments in a Perl script are given by #.

for example

```
# This is a factorial program
# It is written by A. Verma
...
..
```

and then the code, you can give the comments in between the program statements also but they should begin with a #.

9.1.7 Array

When more than one variable of the same type needs to be declared to capture some vector values, then we declare arrays as in many other programming languages.

The arrays do not need any specific declaration to be created, they are also created the first time they are referred just as other variables in Perl are created the first time they are referred.

The only difference is that when you refer to the array the first time, we need an @ character to create it and then the normal $ character is used to access each element of the array.

for example

```
@arr[2] = 6;
```

This will create an array by the name of "arr" an store at its 3rd element a value of 6, since just as in C language the first array element begins with a subscript of 0.

Then we can refer to other elements of the array as:

```
$arr[10] = 55;
$arr[11] = 50;
```

If we wish to refer to the whole array, then we will use @arr.

To print the first 10 elements of an array, we can write:

```
for ( $a = 10; $a < 10; $a++)
{
print("array element is $arr[$a] \n");
}
```

9.1.8 Hash

A hash is a special type of data structure in Perl which is similar to an array but differs in the fact that instead of giving numeric values as array subscripts , we can give string values also in as hash.

For example,

```
$nucleotide{"A"} = 1;
$nucleotide{"C"} = 2;
$nucleotide{"G"} = 3;
$nucleotide{"T"} = 4;
```

In hash, we have a curly bracket ({ }) instead of square brackets ([ ]) for subscripts as in arrays.

Also to refer to the whole hash at once we use the % sign instead of the @ sign as in arrays of Perl language.

9.1.9 Strings

When anything is enclosed in double quotes in Perl, it is a string.

For example,

```
print(" x is equal to $x \n");
```

The pint function uses a string as an argument.

Suppose we have

```
$str1 = "may"
$str2 = "God"
$str3 = "bless"
$str4 = "you"
$str5 = "$str1 $str2 $str3 $str4";
```

Will give you the value of str5 variable as:

"may God bless you"

If you don't want that the values of the variables should be evaluated then you will have to enclose in single quotes.

$str5 = '$str1 $str2 $str3 $str4';

The value of the str5 variable in this case will be as follows:

"$str1 $str2 $str3 $str4"

Suppose you want to add some character to a string without spaces then if you give

$str1 = "love";

$str2 = "$str1 ly";

This will produce the value of str2 variable as

"love ly"

To make it one word without spaces, you have to give

$str2 = "${str1}ly";

This will now have the value of str2 variable as

"lovely"

If you want double quotes to be included as part of the string, then you will have to use the escape sequence \ (backslash).

For example,

This happened by the grace of "God"

You will have to write

print("This happened by the grace of \"God\" ");

9.1.10 Functions

Functions are written so that if we have some code which needs to be executed again and again. Then it is written at one place called the function and that function is called from the main program whenever we need to execute this same piece of code with different values. The values are passed as arguments to the function. It is not mandatory to pass values to the function, there may be some function which just execute the code and do not require any arguments.

Suppose we have a program which takes 5 values and simply prints those 5 values in the function. We call this function twice each with different set of 5 values which are passed as parameters.

For example,

#Declare an array of 5 elements

@a1 = (2, 4, 6, 8, 10);

#Declare another array of 5 elements

```
@a2 = (1, 3, 5, 7, 9);
#Call the print function with 1st array elements
print("The first array is \n");
print_values( $a1[0], $a1[1], $a1[2], $a1[3], $a1[4]);
#Call the print function with 2nd array elements
print("The second array is \n");
print_values( $a2[0], $a2[1], $a2[2], $a2[3], $a2[4]);
#Exit the program
exit;
#The function print_values
sub print_values
{
    my $a = $_[0];
    my $b = $_[1];
    my $c = $_[2];
    my $d = $_[3];
    my $e = $_[4];

    print("1st element is $a \n");
    print("2nd element is $b \n");
    print("3rd element is $c \n");
    print("4th element is $d \n");
    print("5th element is $e \n");
}
```

The output on the screen will be

The first array is

1st element is 2

2nd element is 4

3rd element is 6

4th element is 8

5th element is 10

The second array is

1st element is 1

2nd element is 3

3rd element is 5

4th element is 7

5th element is 9

Here the subroutine or function print_values is called twice from the main program with different set of values and it prints those values on the screen.

In the sub routine print_values, the my keyword declares the variable $a, $b, $c, $d, $e as local variables, so they are available only to this function and their values changed in this function will not be affected anywhere else.

The arguments which are passed to the function print_values are stored in a special array _(underscore) and its elements $_[0] is the 1st argument that is passed to the function is stored in the local variable $a and so on.

Suppose you want to write a program which multiplies 2 numbers and returns the answer.

```
$x[0] = 2;
$x[1] =3;
$answer = multiply( $x[0], $x[1]);
print("The first set of numbers multiplied is $answer \n");
$y[0] = 5;
$y[1] =5;
$answer = multiply( $y[0], $y[1]);
print("The second set of numbers multiplied is $answer \n");
exit;
sub multiply
{
        my $a = $_[0];
        my $b = $_[1];
        my $ans = $_[0] * $_[1];
        return ($ans);
}
```

Here again we call the function multiply twice with 2 different set of values. In multiply function we use a local variable $ans to compute the result and pass it back to the main function by the return statement. In the main program, the value returned from the subroutine is captured in the $answer variable and that is printed.

The output looks like

The first set of numbers multiplied is 6

The second set of numbers multiplied is 25

9.1.11 Input and Output

Files are used in Perl for reading and writing.

The files can be opened with an open function. They can be opened for

1. reading only given by the character "<" ,
2. writing only given by the character ">"
3. appending (i.e begin writing from the end of file and do not overwrite) given by the character ">>",
4. read and write both given by the character '+>"

Just like in C language, a file pointer is associated when we open a file and the file pointer is used for referring to the file later.

Consider the following piece of code :

```
open( FILE1, "< abc.txt");
while (<FILE1>)
{
 $p = $_;
 print($p);
}
```

Here "abc.txt" is the file to be opened. FILE1 is the file pointer and the "<" character used in "<abc.txt" is showing that the file should be opened in read only mode. The while loop checks the position of the file pointer. The "< >" operator is used to read the line from the file, until it reaches the end of the file.

The line of the file read is stored in a special variable $_. In our program, we copy it into our local variable $p and then use the print function to print the line read on the screen. The while loop operates till there are no further lines in the file.

9.2 BIOPERL

http://bioperl.org is the site where more than 500 Perl modules are kept for the purpose of helping the bioinformaticians to use the bioinformatic databases and write small programs in the form of scripts to get useful information out of the databases.

BioPerl is a large project which is maintained by a group of volunteers and it is open source. Open source means that the code is available to everyone for debugging and suggestions are invited to modify it, you can test your programs on the code.

The programming language that is used in BioPerl is basically set of Perl Commands which incorporate object oriented programming concepts in it.

BioPerl project has now taken a shape where useful code and correct documents are available for bioinformatic source of information.

The documentation of the code is available in bptutorial.pl which comes along with the BioPerl package.

BioPerl can be used with any of the operating systems such as Unix /Linux , Windows and Mac. But most commonly it is used on the Unix/Linux Platform.

To install BioPerl you will have to read the INSTALL file that comes with the BioPerl documentation.

In Unix to install BioPerl you will have to download the tar file from the website and you will need to untar it, then you will need to do configure and make processes to complete successful installation of the file.

9.2.1 Methods

Procedures or functions which are called to get different objects are part of the BioPerl package. A few are described here:

1. get_sequence

   This is used to get a sequence from a database. This requires the internet to be connected.

   my $seq-obj = get_sequence ('swissprot', "ROA1-HUMAN");

   you can also get a sequence from other databases with accession number:

   $seq_obj = get_sequence ('genbank', "AI144021");

2. read_sequence

   This is used to read a sequence from a file and put it into an object.

   $seq_obj = read_sequence ($file1);

   The above will take a format on its own by looking at the extension of the file.

   $seq_obj = read_sequence ($file1, 'swissprot');

   The above will fetch the sequence in the format which is used in swissprot database.

3. read_all_sequences

   This reads all the sequences in the file in an array of objects.

   @seq_all_obj = read_all_sequences ($file1, 'genbank');

4. write_sequence

   write_sequence(">$file1", 'genbank', $seq_obj);

   This will take the sequence from the $seq_obj object and write it onto the file file1 in the genbank format and it is opened in write mode.

5. new_sequence

   You can create a new sequence by giving the nucleotides of the sequence in the string, followed by the sequence name and followed by accession number and store it into an object at the same time by giving as follows:

   $seq_obj = new_sequence ("ATGTTATCTGTATGC", "newname", "AI120091");

6. print

   You can use the print statement to print the attributes of an object as follows:
   print "Sequence Name:", $seq_obj->display_id, "\n";

   print "Sequence Accession No:", $seq_obj->accession_number, "\n";

   print "First 10 Nucleotides are:", $seq_obj->subseq(1,10), "\n";
   This will print the output as follows:

   Sequence Name: oldhuman

   Sequence Accession No:AI146529

   First 10 Nucleotides are:ATCCGTACGT

7. To get the whole sequence stored in the sequence object.

   $seq_str = $seq_obj ->seq();

   This will fetch the whole sequence stored in the seq_obj into the new variable $seq_str.

   Note that $seq_obj belongs to the Bio::Seq objects which will be described shortly.

8. blast_sequence

   To connect the sequence into a BLAST format we have to use the blast_sequence function as follows;

   $blast_view = blast_sequence ($seq_obj);

   This will put the BLAST result in the newly created variable $blast_view.

9. write_blast

   To view the BLAST result of the sequence which was put to BLAST in the previous example can be done with the help of write_blast function as follows:
   write_blast(">new.blast", $blast_view);

   Now the content of the object $blast_view will be stored in a file called new.blast whose contents can be viewed with the help of cat command as follows:

   $cat new.blast

9.2.2 Objects in Bioperl

As the BioPerl project uses the concepts of object oriented programming. There are a number of objects which are defined in BioPerl to be used for programming. These have members and methods both defined in the base objects which can help in making the program easy to code.

1. Bio::Seq

   This is the main sequence object. The one we defined previously as $seq_obj was belonging to this object.

2. Bio::PrimarySeq

   This is the sequence object without features.

3. Bio::SeqIO

   This is used for doing file input and output with sequence.

4. Bio::DB::GenBank

   This gives access to the GenBank Database. Similarly other databases of bioinformatics can also be accessed by giving their names instead of GenBank.

5. Bio::Tools::Run::StandAloneBLAST

   This runs BLAST program on your local computer.

6. Bio::SimpleAlign

   This is used to deal with simple Multiple Sequence Alignments.

7. Bio::UnivAlign

   This is used to deal with multiple sequence alignments.

8. Bio::Tools::pSW

   This aligns the 2 sequnces od Smith Waterman Algorithm

9. Bio::AlignIO

   This aligns 2 sequences with BLAST.

10. Bio::SeqFeature

    This is the sequence feature object used in BioPerl

11. Bio::Tools::RestrictionEnzyme

    This is used to find out the restriction sites in the sequences.

12. Bio::Tools::Sigcleave

    This is used to find out sites where amino acids can be cleaved or in other words where amino acids can be broken.

13. Bio::Tools::Seqpattern

    This helps in describing the sequence patterns of regular expressions.

14. Bio::Tools::Genscan

    This helps in finding Genes.

When you require these objects in your program, then at the beginning you have to give declarations like:

use Bio::Seq

use Bio::SeqIO

use Bio::SeqFeature

use Bio::Tools::Seqpattern

You can use a few other methods of sequence objects to find other attributes.

Suppose that an object $sequenceobject has been populated with some data from somewhere.

You can use the following methods to gather more information about its attributes.

1. $sequenceobject-> top_seqfeatures

   This returns the top level sequence features of the object Bio::Seqfeature

2. $sequenceobject->all_seqfeatures

   This returns all the sequence features including sub sequence features.

The following methods return news sequence objects.

Suppose we want to truncate from 2 to 7 positions of nucleotides and produce a new object. Then if we write

print "New Truncated Object is:", $sequenceobject->trunk(2,7)->seq(), "\n";

This will produce the output on the screen as :

New Truncated Object is: TACTAG

if the original sequence of the object had TACTAG from positions 2 to 7.

9.3 LINUX OPERATING SYSTEM

Linux Operating System is by enlarge the most accepted operating system among Bioinformaticians as compared to MAC and Windows Operating Systems because of the following reasons:

1. Many good scientific software have been specifically written for Linux /Unix.
2. Unix/Linux is able to configure and install tools which are helpful in Bioinformatics such as Perl, Python.
3. It has command line interface, where commands are written and executed easily.
4. Linux/Unix supports pipelines.
5. Shell scripts can be easily written for programming and execution.
6. Many pre-compiled binaries are available in Linux/Unix.
7. It supports multiuser and multitasking systems.

Some Basic Commands in LINUX

1. pwd : present working directory

   $pwd

   This command when given at the shell prompt results in telling in which directory are you working currently.

   For example, in the above the $ is the shell prompt and the command given afterwards will result in the following:

   $pwd

   /usr/archana/newdir

   $

Where / is the root directory and underneath the / is the usr subdirectory which is the standard directory of the Unix file system where all users work. Underneath the /usr is the archana subdirectory which is the directory name of the person who has logged in, which is again made for each user by his username under the standard /usr directory. Underneath /usr/archna is a directory newdir which the user may himself would have created. So presently the user is working in the directory newdir.

2. cd: change directory

   This command when given on the command line changes the directory or takes you to another present working directory.

   Suppose you give the following set of commands on the shell prompt:

```
$pwd
/usr/archana/newdir
$cd ...
$pwd
/usr/archana
$
```

   Here your present working directory was /usr/archana/newdir earlier, then you gave the change directory command with double dots (..) which stands for the parent directory, so you instructed to change to the parent directory and when you gave the command pwd next, it shows you have moved to the parent directory of newdir and now your present working directory becomes /usr/archana.

3. mkdir: create a new directory

   The command is used to create a new directory underneath the present working directory.

   Suppose you give the following set of commands:

```
$pwd
/usr/archana/newdir
$mkdir dir1
$cd dir1
$pwd
/usr/archana/newdir/dir1
```

   Earlier you were in the present working directory /usr/archana/newdir, you made a new directory underneath newdir by the name of dir1. Then you changed to the new directory dir1 and when you gave the pwd command next, it shows that the new directory has been created and that is now your present working directory.

   /usr/archana/newdir/dir1

4. rmdir: remove a directory

   This removes or deletes a directory. To remove a directory, first it should not have any files existing in it and your present working directory should be its parent directory to remove that directory. You cannot be in that directory and remove it.

   Follow the following set of commands:

```
$pwd
/usr/archana/newdir/dir1
$cd ..
$rmdir dir1
$pwd
/usr/archana/newdir
$cd dir1
error: directory does not exist
```

   Earlier you were in /usr/archana/newdir/dir1 directory, you changed to the parent directory of dir1 which was newdir. Then you removed dir1 directory. You checked where you are now by giving the pwd command. Then you tried to get into dir1 directory again and it gave an error that the directory does not exist since you had just removed it.

5. ls: listing of files and directories

   This command would only give the names of the files and directories present in the directory you are currently working in without giving other details of the files.

```
$pwd
/usr/archana/newdir
$ls
file1.c
file2.c
abc.txt
$
```

   This will give just the names of the files and directories present in newdir.

6. rm: remove a file

   Just as we had removed a directory, we can also remove a file from the directory we are currently in.

   Follow the following set of commands:

```
$pwd
/usr/archana/newdir
$ls
file1.c
```

```
file2.c
abc.txt
$ rm abc.txt
```

Do you want to remove abc.txt (y/n)

```
y
$ls
file1.c
file2.c
$
```

Here you were in the newdir directory which had 3 files. You removed abc.txt and if asked for confirmation since you may have accidently given the command. You said yes by giving the y option. Then you listed again and only 2 files were left, abc.txt was gone.

7. grep: fetches the line that contains the pattern in the specified file.

   Suppose the file contains the following contents:

   file1.txt

   Welcome to Unix

   It is a very popular operating system.

   Linux is the graphical edition of Unix.

   It is a multitasking operating system.

   If we gave the following command on the command line

   ```
   $grep Unix file1.txt
   Welcome to Unix
   Linux is the graphical edition of Unix.
   $
   ```

   The grep command fetches the lines that contain the word Unix and does not fetch those lines which do not contain the word or pattern Unix in the file file1.txt.

8. touch: changes the time that the file was last accessed or modified.

   The ls –l command is the same ls command with the –l option of long listing. This gives all the details of the files present.

   For example if a directory contains 3 files

   ```
   $ls
   file1.c
   file2.c
   abc.txt
   $
   ```

Then giving the ls –l command will result in the following:

```
$ ls –l
-r w x - - - r - - geeta public 148 Nov 7 10:30 file1.c
-r - - r w - r - - sita public 36 Jun 10 11:40 file2.c
-rwx- w x r - x pita public 220 Apr 4 3:20 abc.txt
$
```

Here the first – (hyphen) indicates that it is a file. The next 9 characters give the file permissions which will be explained shortly. The next (geeta, sita, pita) in this case give the name of the owner and public gives the name of the group to which the file belongs. The next integer gives the number of bytes of the file, then it the gives a timestamp that the file was last accessed or modified and finally the name of the file.

The touch command simply takes the system time and date and attaches that to the file's timestamp and date deleting the older timestamp.

For example,

If I now give

```
$touch abc.txt
$ls –l
-r w x - - - r - - geeta public 148 Nov 7 10:30 file1.c
-r - - r w - r - - sita public 36 Jun 10 11:40 file2.c
-r w x - w x r - x pita public 220 Oct 29 6:44 abc.txt
```

Here the timestamp of the file abc.txt has changed assuming that the current date and time is Oct 29 6:44.

9.3.1 File Permissions

The strings r w x - - - r - - , r - - r w - r - - and r w x - w x r - x above indicate the different file permissions for the 3 files respectively. The organization of the 9 characters is as follows:

rwx	rwx	rwx
owner	group	others

The 9 characters are divided into 3 groups of 3 characters each. Each group always has the read then write and then the execute permission character. The first group is the owner of the file, it contains the read write and execute permission given to the owner of the file. The next group is the group to which the owner belongs, all group members of the owner's group would have the next read write and execute permission in that order. The last division is of the rest of the world for that file, people who are neither the owner of the file nor do they belong to the group of the owner and are called others. Their read write and execute permissions are given in that order.

For example, consider the following:

```
$ls –l
-r w x r - x - - x geeta public 148 Nov 7 10:30 file1.c
```

Here the owner of the file is geeta and she has the first 3 permissions of read write and execute to her file file1.c. geeta is a member of the public group, all members belonging to geeta's group which is public have read and the execute permissions. That is they can read or run the file but can't make changes to it. The other people only have the execute option that is they can neither read or write to the file simply run and see the output.

9.3.2 Redirecting The Output

The redirect output symbol (>) instructs the shell to redirect a command's output to the specified file instead of the terminal.

For example,

```
$cat file1.c file2.c > newfile.c
$
```

Here the cat command is the concatenate command which joins two files file1.c and file2.c and sends the joined file's output to the newfile.c

9.3.3 Piping

The Unix pipe facility lets you connect programs to other programs. In other words, the shell uses a pipe to connect the standard output of one command directly to the standard input of another command.

It does this with the help of a pipe symbol (|) between the two commands.

For example,

```
$grep Unix newfile.c | sort
```

Here we are using the grep command as the first command. It is supplied with two arguments, the first the pattern to be searched and the next the file in which the search is to be made. The second command is the sort command.

The output of the grep command becomes the input for the sort command. Then the sort command sorts whatever is given to it and prints on the screen.

9.3.4 Mounting Unmounting Files

Mounting basically means linking the file system on the floppy drive to the main file system of Unix which is the root file system present on the hard disk. Once mounted, the files and directories can be created on the new file system and treated as normal files or directories existing in the main file system.

The mounted file system behaves like any other file system, having a root directory and other subdirectories branching out of it.

The actual meaning of mounting is simply attaching the root directory of that file system to any point in the root file system. This point of attachment is called the mount point for that file system. The mount point of any new file system must always be specified as a path from the root directory of the existing file system. The default mount point is /mnt.

$ mount /dev/fd096ds15 /mnt

Here mount is the command and /dev/fd096ds15 is the floppy drive which is to be attached in the root file system at the mount point given by /mnt.

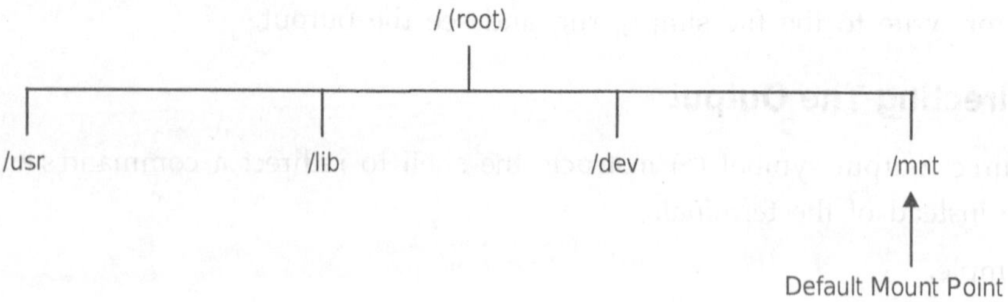

Suppose /dev/fd096ds15 contains the following directories and files:

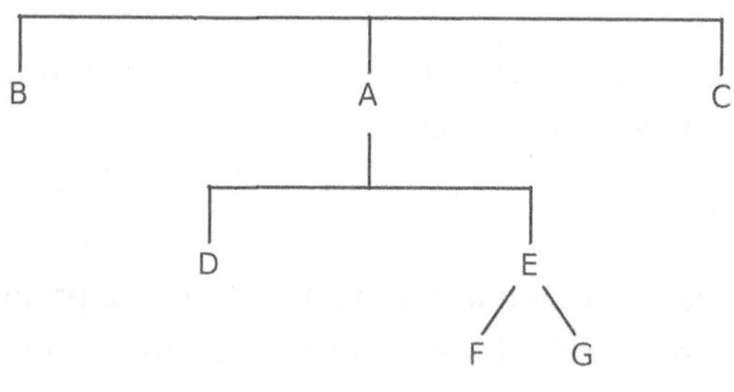

Then after the mount command we will have the file system arranged as follows:

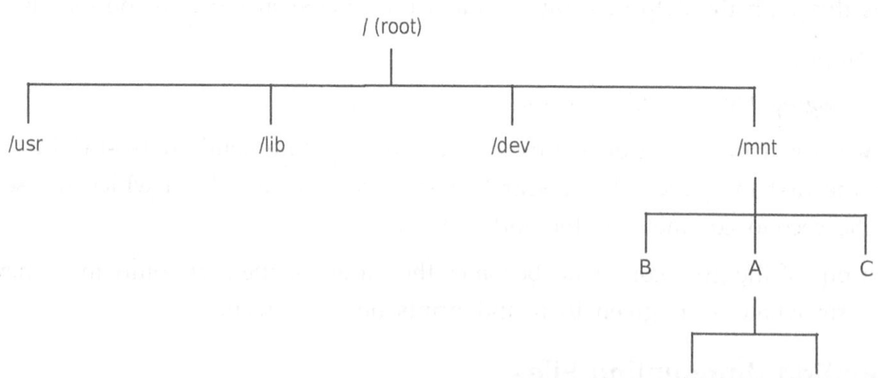

We can also make any other mount point instead of the default mount point by giving the following command:

$mount /dev/fd096ds15 /usr/oracle

Here the mount point becomes /usr/oracle, so the file system looks like this:

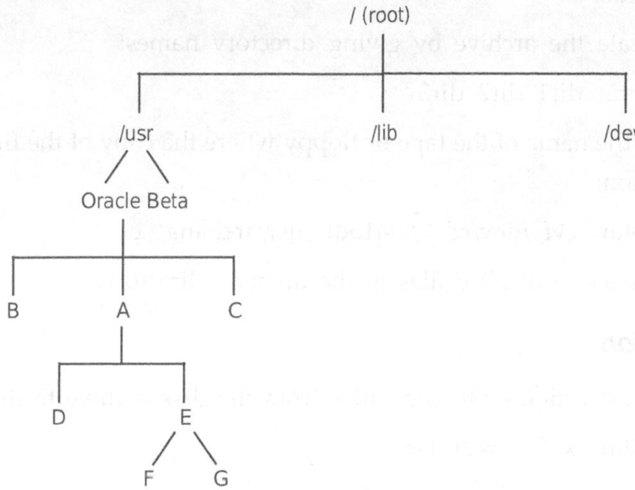

9.3.5 Unmounting a File System

Unmounting a file system, means detaching the new file system from the root file system. On unmounting, the files on the file system remain undisturbed, only thing that happens is that they become inaccessible. To access the files once again, the file system has to be remounted using the mount command. To unmount, we use the umount command and give the mount point as parameter, i.e. the mount point at which the new file system is mounted.

```
$umount /mnt
```

Or

```
$umount /usr/oracle
```

in the above case.

9.3.6 tar

The tar command is used to copy files or directories on the disk archive and restore back from the disk to the current directory.

It has 4 options:

-x - copy from disk archive to current directory

-c - copy from current directory to disk archive

-f - to create a disk archive

-v - to display the files in the disk archive while tar is in progress

Suppose we use the following command:

```
$tar -cvf new.tar file1.c abc.txt
a file1.c 360k
a abc.txt 24k
```

The .tar extension is used in the archive name so that later you remember it was created from the tar command.

You can also create the archive by giving directory names:

$tar -cvf newdir.tar dir1 dir2 dir3

We can also give the name of the tape or floppy where the copy of the files have to be made by the following option:

$tar -cvf /dev/rdsk/f0q15dt /usr/archana/*.c

This makes the backup of all c files in the archana directory.

9.3.6.1 The –x option

This is used for extraction or to copy files from the disk archive to the current directory.

$tar -xvf newdir.tar

Or we can give

$tar -xvf /dev/rdsk/f0q15dt

x /usr/archana/file1.c 24 bytes 1 tape block

x /usr/archana/file2.c 220 bytes 1 tape block

9.3.6.2 The -t option

This is used for displaying the files in the archive.

$tar -tvf new.tar

-rw- r - - r - - 105/2 34568 Aug 29 17:55 2002 file1.c

-rwx r - x r - - 105/2 2122 Sep 30 18:50 2008 abc.txt

Or we can give

$tar -tvf /dev/rdsk/f0q15dt

-rwx r - x r - - 201/10 322 Jul 10 06:45 1990 ./file1.c

-rwx r – x r w x 201/10 644 Jul 10 07:48 1990 ./file2.c

9.3.7 gzp/gunzip

This command is used to compress or decompress the archive created by the tar command.

The command to compress the archive is:

$gzip new.tar

This has created a zipped file of the tar file by the name of

new.tar.gz

This can now be sent as an attachment or mail or to do the ftp (file transfer protocol)

9.3.7.1 gunzip

Before we use the -x option of the tar command, we may like to decompress the zipped or compressed tar file by the gunzip command as follows:

```
$gunzip new.tar.gz
$tar -xvf new.tar
```

9.3.8 Telnet

Telnet allows a user at one terminal to access the terminal at another site which is at a remote place by giving the IP address of the remote machine.

The commands are run on the remote machine, you type from your keyboard and terminal but actually those commands are run on the remote machine at another site and its answers are displayed here.

For example,

Type telnet on your Unix system's command prompt or shell prompt as follows:

```
$telnet 192.215.13.4
```

where the numbers show the IP address or the (internet protocol) address of the remote machine to which you want to connect.

```
$telnet          192.215.13.4
Trying           192.215.13.4
Connected to     192.215.13.4
Login:
Password:
telnet>
```

Once the remote machine is connected, a login password appears for that machine, you should have been given a login password by the system administrator of that machine to access it. Once you are through with the login process, a telnet prompt appears on which you enter the commands which you want to execute on that machine and their output will be shown on your terminal.

When you are at telnet prompt, to log out of that machine use quit command.

```
telnet> quit
$
```

And you will go back to the Unix shell of your original machine by the sign of the dollar prompt.

9.3.9 ftp

The full form of ftp is the File Transfer Protocol. This command is used to transfer the files between two machines at different sites.

The ftp protocol can use either the IP address of the remote machine or the user name of the host machine as the second argument to connect to the machine at the remote site as follows:

```
$ftp 192.215.13.4
      OR
$ftp goodies
Connected to goodies
Name:
Password:
ftp>
```

Once you are logged in, the ftp prompt will appear. ftp accepts all commands related to files and directories on the remote machine such as:

```
ftp>pwd
/home/geeta
ftp> ls -l
-r w - r - - .....
- r w - r - - ......
- r w x r - - ....
ftp> mkdir
ftp>rmdir
ftp>delete file1.c
ftp> mdelete file1.c file2.c file3.c
```

delete is the command used on ftp to delete a remote file and mdelete is used to delete multiple files.

The files can be transferred back and fro by the get and the put commands of ftp as follows:

```
ftp> put peacock.pdf
local: peacock.pdf remote: peacock.pdf
200 PORT command OK
150 opening ASCII mode data connection for peacock.pdf
226 transfer complete
3148 bytes sent in 0.02 seconds (150 Kbytes/s)
ftp>
```

Now here we have put a local file peacock.pdf on the remote machine by the same name and it shows it is successful and gives us back the ftp prompt.

```
ftp> get /home/geeta/newfile.tar herefile
200 PORT Command OK
```

```
150 Opening ASCII mode data connection for newfile.tar (4418 bytes)
226 Transfer Complete
4418 bytes received in 0.03 seconds (150 Kbytes/s)
ftp>
```

here a file is brought from the host machine from /home/geeta directory by the name of newfile.tar and it will be placed in the local directory herefile.

Then to close the connection of ftp we type

```
ftp>close
221 GoodBye
ftp> quit
$
```

First we close the connection by close command and to return to our machine we type quit and get back the $ prompt of our machine.

9.3.10 Developing Applications on Linux OS

There are two ways in which you can run applications on the LINUX OS. One way is to get the freely available binaries on the web and install on your Linux system and run them. The second way is to develop Unix shell scripts to run whatever programs you want to see the output. Let us discuss each separately.

1. Taking Binaries from web: Software on web is normally available in the form of tar files. The tar files can be downloaded and decompressed into the local subdirectory where you want it.

The software on the web may be available either in the form of source code which you would need to compile and make it into an executable file or in the form of executable files only known as binaries.

For example, certain binaries available on the web are in the form of

```
blast.Linux.tar.z
blast.Solaris.tar.z
blastZ.exe
```

which can be downloaded. The instructions are generally provided in the README or INSTALL file which you will need to read and follow the instructions step by step. You would need to create a subdirectory which is local where you can move the binaries. Then you can uncompress them. If there are any configuration or installation scripts, then you would need to run them and finally link the binary into a directory which is in your default path and run the binary file eventually.

If the binaries are not available on the web, then there would be source code available for which you will have to compile the source code by using the make command of the Unix system.

Normally the source code would include the header files with the .h extension and some files with the .c extension.

A makefile will tell how to put all files together to make a single executable. Normally a makefile contains a number of compile commands such as:

```
$cat makefile
shape :          width.o breadth.o
                 cc -o shape width.o breadth.o
width.o :        width.c formation.h
                 cc -c width.c
breadth.o :      breadth.c formation.h
                 cc -c breadth.c
formation.h :    heights.h line.h
                 cat heights.h line.h > formation.h
$
```

Here the make file will be executed from bottom to top to make the final executable file called shape.

When we run the make command on the Unix prompt, then the commands written in the makefile are executed. For example for the above makefile, they will be executed as follows:

```
$make
cc -c breadth.c
cc -c width.c
cc -o shape  width.o  breadth.o
$
```

cc -c option tells to make an object file with the .o extension of the source file with the .c extension.

cc -o option tells to make the object file with the name given next, in this case it is shape. Again when your executable is made you may link it with ln command of Unix system to set in your default path and run the executable.

2. Unix Shell Scripts: Unix shell scripts are a number of Unix commands which are written in some order to form a small program which is called the shell script and has an extension of .sh.

For example let us see the contents of a shellscript named myscript.sh with the help of the cat command of Unix system.

```
$cat myscript.sh
    foreach file ($*)
    echo "The name of the file is $file"
```

```
gb2fasta $file > $file.new
formatdb "$file" $file.new
end
$
```

In this script we have a variable called file which takes each file given on the command line. Then the echo command prints the name of the file on the standard output.

The gb2fasta is a command which converts the GenBank sequences to FASTA format and sends the output to a new file with the .new extension instead of the standard output by the redirection symbol (>). Then the formatdb command processes the new files in a format on which the BLAST can then work on.

Now once our shell script has been made, we need to provide it the executable permission wit te help of chmod(change mode command) of the Unix system.

```
$chmod a + x myscript.sh
```

In a+x, a means all(owner, group and others) are given the permission with the help of + sign. The executable permission shown by (x) to the file myscript.sh.

So if we do

```
$ ls -l myscript.sh
-- - x - - x - - x ........... myscript.sh
```

If it does not have any other read or write permission. Now that we have given the executable permission we need to run the file.

We can do so by

```
$./myscript.sh *gb*.seq
```

The .(dot) says that run from the current directory. The ./ says run the executable myscript.sh under the current directory and provide the command line arguments *gb*.seq. which says all files containing gb (for GenBank) in the filename anywhere with 0 or more characters on the left and 0 or more characters on the right of the filename having the extension of .seq.

In the program these files will be available in the ($*) of the shell script, one by one and the file variable in the shell script will execute on each one of them separately by the for each command and run gb2fasta and formatdb commands on them as given in the shell script.

9.4 UNDERSTANDING AND USING BIOLOGICAL DATABASES

There are a number of biological databases which we have discussed in the earlier chapters.

You can either access the biological databases by taking the type of data that is stored in them or you can deposit your own sequences in the open source databases.

For an example, let us take GenBank and PDB (Protein Data Bank). As you know GenBank is mainly used for storing DNA sequences and PDB is used for storing molecular structure data.

The different databases can be accessed at their sites.

GenBank for example contains datatypes as follows:

1. mRNA
2. cDNA
3. Genomic DNA
4. ESTs

These terms have been explained earlier. You can search the GenBank database by a text based query search or you can use the BLAST program to compare a query DNA or Protein Sequence to the sequences which are already stored in GenBank.

There are different interfaces to the search.

9.4.1 Downloading Sequences

You can download the different types of data from these public databases in the file formats such as :

1. FASTA format
2. GenBank flat file format
3. ASN.1 (Abstract Syntax Notation one format)

When you want to deposit the data then the data entered by user is converted into an appropriate format which is suitable for the database.

After downloading, you can save the results in the appropriate formats.

FASTA is the most common format in which you save your files.

The formats can be converted into other formats by the user interface provided.

To deal with multiple, complex and comprehensive data, Entrez tool is used to deal with such type of data which we have discussed in earlier chapters.

9.4.2 PDB

The PDB data is available in 2 formats:

1. PDB – Flat File Format
2. mmCIF Format

There are different tools used for validation of data before we can deposit the data into the database.

9.5 OVERVIEW OF JAVA

Java has developed many applications which are easy to access and available over a large number of platforms and also provides computations which can be made at different levels of difficulty.

The most common and popular and the oldest language for Bioinformaticians has been Perl. It is because of its capability to answer direct questions easily with its programming capability which includes regular expressions and bioinformatics modules based programming. Also Perl has the capacity to transfer data over the internet using modules.

The problem with Perl scripts is that they are dependent on some prerequisites for installation and do not support Graphical User Interface.

The advantage of Java lies in its Graphical User Interface support and its capability to provide programming at different levels of computations so that different types of work can be carried out.

There are different types of Java APIs which are specifically designed for bioinformatics and access the bioinformatics databases to retrieve the information.

For example EnsEMBL is a genome browser, for comparisons, they need to access the genome data. Then we can give the organisms name to these browsers and get its sequences. EnsEMBL uses different types of Java APIs to retrieve data.

To perform analysis of bioinformatics data, the types of tools that are used are as follows:

There are a number of applications which use Java for analyzing the data.

1. TIGR Multi Experiment Viewer (MEV) – Analyses Patterns of gene expression. It is open source.
2. J – Express is a commercial Java application. This provides microarray/genechip analysis.
3. WebMol uses Java 3D to analyze and visualize molecular structures of complex proteins. Here Java Applets are used.
4. BioJava is an API rather than an application. This manipulates biological data and does transfer of data also.
5. Apollo – This is a Java Bioinformatics tool which helps in writing explanations or footnotes about genes in various genomes. Here Human Facilitated annotation activities can also be performed.
6. Sockeye – This also uses Java 3D to compare annotations of several genome sequences at the same time.

9.5.1 BioJava

BioJava processes biological data wit the help of Java tools and it is an open source project.

BioJava basically contains a set of library functions which are written in the Java programming language for the purpose of manipulating sequences and protein structures.

The types of data the BioJava supports includes DNA and protein sequences as well as 3D protein structures.

Much of the everyday tasks which require a lot of time and are boring are carried out automatically using BioJava such as parsing of PDB files.

The API of BioJava ranges from providing file parsers, data models and algorithms to convert the data into standard formats and to help in analyzing the data.

The main functions that BioJava can do are:

1. Access protein and DNA sequences from remote and local databases.
2. Change formats of files and records.
3. To see the protein structures and to manipulate them.
4. To manipulate sequences.
5. Finding sequences which are similar.
6. To check sequence alignments.

BioJava is a module driven project in which the Java classes are used to model amino acid or DNA sequences. The names of classes are meaningfully kept so that it is easier for the biologists to identify which class is for which purpose.

Some algorithms are used to do protein comparisons and systematic comparisons of all proteins in the PDB database on weekly basis.

The genome and sequencing modules create gene sequence objects from the main module and these objects can then be written in different formats and imported into databases.

There is a module for pairwise sequence alignment in which Needleman and Wunsch and Smith and Watermans algorithms have been incorporated to give the outputs in various formats.

9.6 CORBA

The size and complexity of Bioinformatics data has been increasing considerably. So the distribution of bioinformatics data becomes mandatory.

CORBA is a standard which tries to solve the problem of cross-platforms. It tries to hide the underlying hardware platforms, operating systems, databases and programmimg languages to bring everything closer under one common platform.

Most common form of biological data is in the form of flat-files and the SRS or the Sequence Retrieval System puts the data in terms of flat files to transfer it.

Flat files have their disadvantages in terms of not being able to keep the information of one entry together.

To overcome the flat file system is dependent on removing their formats. CORBA provides an interface to call methods either locally or remotely without telling the difference between the two.

The Object Management Group's (OMG) primary goal is to standardize and promote object technology and it uses CORBA(Common Object Request Broker Architecture) as its Core standard.

CORBA uses the concept of interfaces combined with distributed object oriented programming techniques.

In addition it defines:

1. IDL (Interface Definition Language) which is language independent and provides the description of objects which are public.
2. ORB (Object Request Broker) which gets the requests from the clients and without your knowing any details about how things are done, would put the requests in objects.

The IDL defines an interface for a distributed object, any client having public access to the object can use it. The object language and whether it is located locally or remotely is of no concern. The client needs reference of the object to invoke methods on the CORBA object.

CORBA invocation is similar to Remote Procedure Call (RPC). In RPC a client invokes a function on the server, whereas in CORBA the client invokes a method on the object.

The interfaces of the IDL represent every class of the data model and every database entry is represented as a CORBA object.

The queries in CORBA are standardized. Again a CORBA object acts as a query evaluator. The query evaluator takes a query string as its argument and returns a reference to the collecton of results.

The results returned are again in the form of a CORBA object.

The query evaluator can be used with any type of query language and any type of its underlying database.

CORBA is a model driven architecture which has an independent platform model and a dependent platform model based on its operating system. A mapping is made of the dependent platform model to the independent platform model to hide the underlying network etc.

CORBA brings heterogeneous distributed systems together to work under one roof.

CORBA's main aim is to enable interoperation between distributed bioinformatics resources.

Many bioinformatics databases have provided CORBA servers to access information.

9.7 XML

XML is an open standard which is programming language and operating system independent.

XML stands for Extensible Markup Language which is used to construct structured documents.

In the context of bioinformatics, XML is able to clearly indicate portions of documents which are meant for raw sequence data, or accession number or sequence identifiers or references to public databases. It makes it easy for the end user to extract the piece of information he is looking for.

For example a record in XML may look like this:

<?xml version = "1.0" encoding = "UTF -8"?>

<sequence>

```
<accession> HU-1546744 </accession>
<organism> Homo Sapiens </organism>
<sequence data>
    acgtgtcagtactttggaacggatcgatagtaagttggccttatagct
    ttaccagggtaggacctagatcgatagcttagcaattgaccagtcgac
    ccttaggatcaggatcaaggataccagatagacaggggatcccagat
    cctaggattaggaagagattagggataccagagtgaccagataggac
      ........
</sequence data>
</sequence>
```

XML is a markup language just like HTML and consists of tags, in the above, the sequence data are sequence are tags.

XML documents can be distributed over the network by using the HTTP protocol, but for that a web service or an XML protocol is required.

DAS (Distributed Annotation System) is one such example. This is used to share genomic data and make predictions and notes about genomic sequences, identify intron exon regions, comparing sequences to already identified sequences etc.

DAS is based on the client server architecture and receives data in the form of XML documents.

Clients connect to DAS server, send queries in the URL format and receive responses in the XML document form.

Each DAS sever can have multiple data sources. The clients are accountable for specifying which data source they want to connect to. Then give a DAS command along with its parameters. The parameters are based on the type of command.

For example to fetch a sequence, we must specify the chromosome number along with the start and stop values of the base pairs.

BSML (Bioinformatic Sequence Markup Language) is another format of XML formats which is used to to encode biological sequences.

This is another open standard for storing and transferring biological sequences, sequence features, notes, biological entities and graphical visuals.

With BSML you can import data in existing data formats such as GenBank, Swiss Prot and EMBL file formats and convert these records into BSML.

XML processors can access DBMS or file system and convert data to XML.

XML is highly flexible, the XML documents are in human readable formats.

XML is internet oriented and has high linking capabilities which prove useful to interconnect databases.

XML provides an open framework for defining standard specifications. This helps in organizing Bioinformatics data which is not standardized.

9.8 WEB DEPLOYMENT CONCEPTS

MatLab can be integrated with commonly used bioinformatics tools such as BioPerl, SOAP -based web services.

MatLab application deployment products helps in integrating MatLab algorithms with the applications of C, C++ and Java.

BioPortal is a user friendly web based GUI that helps in deploying popular bioinformatics applications on large scale cluster and grid computing environments.

Web deployment concepts arise because earlier data was kept at one place and it was clustered there only. Now to make it in distributed form we need to apply deployment concepts.

MolBioLib is a C++ software framework and set of applications which are designed to work with common file formats and data types used both in genomic analysis and general data analysis.

MolBioLib is able to identify causative Single Nucleotide Polymorphism in whole genome sequencing, notice balanced chromosomal reshuffle and compute enrichment of messenger RNAs (mRNAs). It performs a variety of analysis tasks such as annotating genomic intervals.

Galaxy is a platform or a tool for deploying distributed computing clusters using the concepts of cloud computing to deploy bioinformatics workflows.

Grid computing has the power of supercomputers to address the complex and large scale bioinformatics problems.

A computational grid is a set of scattered and heterogeneous computing nodes which can share large scale computational facts and added resources beyond boundaries.

A grid is able to solve massive computational problems using numerous computers arranged in the form of clusters and placed in a distributed environment.

INDEX

A

B

C

D

E

P

R

S

T

U

V

W

X